Here you will find short analyses of clusters of important ideas and trends centered on many central themes. It is inevitable that some of these ideas will come to dominate the shape of our future world in the coming decades while others will ultimately fall away as quaint but interesting notions that did not pan out. You will be able to consider this large range of possibilities. And having done so, you will be better prepared for what is to come.

-- John G. Cramer

Emeritus Professor of Physics

University of Washington

The Elegant Future

Works by the Authors:

By Craig Philip Peterson

The Adventures of Jonathon Farmer Series:

JOURNEY TO THE MIDDLE OF THE MAGIC

THE MYSTERY OF THE DARK ARMY

By Craig Philip Peterson and Doug Odell:

THE CATACOMBS OF DOOM

By Elton Elliott, Craig Philip Peterson, and Doug Odell

THE INFINITY ANOMALLY

Anthologies by Elton Elliott:

NANO DREAMS

LIKE WATER FOR QUARKS (with Bruce Taylor)

By Elton Elliott and Doug Odell:

THE PRINCE OF EUROPE, Book One of the Nanoclone Trilogy

THE BISHOP OF ROME, Book Two of the Nanoclone Trilogy

A Quantum Field of Ghosts and Shadows (published as a chapbook and in anthology LIKE WATER FOR QUARKS)

By Elton Elliott and Richard E. Geis:

(Under the Name Richard Elliott)

The Sword of Allah [Fawcett Gold Metal]

The Burnt Lands [Fawcett Gold Metal]

The Master file [Fawcett Gold Metal]

The Einstein Legacy [Fawcett Gold Metal]

THE
ELEGANT
Future

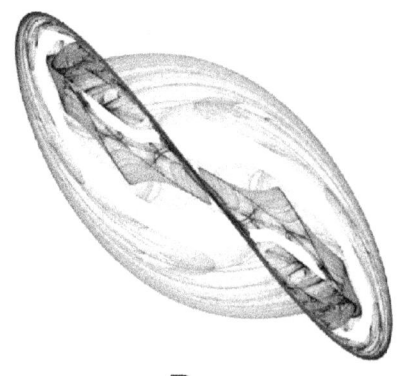

By
Craig Philip Peterson
and
Elton Elliott

Foward by
Dr. John G. Cramer
Emeritus Professor of Physics
University of Washington

THE ELEGANT FUTURE

An MVP Publishing Original.

ISBN 9781072952060

Cover art and interior illustrations by Craig Philip Peterson.

First Printing June 2019.

Printed in the U.S.A.

Dedication:

On February 27, 2019 the authors of this book lost a good and faithful friend, Edward C. Nixon. And while we remember our friend and mourn his loss, the human race lost a true intellectual powerhouse and an advocate for an elegant future.

Acknowledgements:

We thank the following individuals for their help and advice:

Edward C. Nixon

Dr. John Cramer

Andrew Seelye

Doug Odell

Dr. Steve Gillette

Any mistakes are those of the authors to be hopefully corrected in subsequent additions.

Table of Contents

Foreword

Foreseeing the shape of the future through the fog of present possibilities is a difficult and daunting task. Suggestive shapes loom through the obscuring mist, but they can be deceptive or misleading and may dissolve or diminish as they are approached. Other emerging forms, some of great importance, may appear unexpectedly and with no precursors at all.

In 1999, perhaps because I am both an experimental and theoretical physicist with many publications and a science fiction novelist and columnist, I was invited to participate with about 60 others from various backgrounds in an exercise in foreseeing the future. The National Reconnaissance Office, the government outfit that operates our spy satellites, commissioned a prominent DC Beltway think tank to organize and conduct this exercise in predicting what is to come. We participants were divided into groups and given possible future scenarios: the rise of domestic militias that attack civilians and rise to power, digital hacking that precipitates a crash of the worldwide economic system, simultaneous acute shortages of essential industrial materials and fuels, a peaceful exponential rise in world prosperity, etc.

My group was tasked with considering the consequences of a direct attack on the USA by radical middle eastern terrorists. I'm afraid that we blew it. After much deliberation, we confidently concluded that because of the strong moral and civic traditions in the US Military and the Federal Government, our nation would focus on the weakness of the enemy and would not subject our own citizens to intense surveillance or strong anti-terror measures. Essentially, we predicted that developments like Guantanamo, the TSA, the Patriot Act, and the Iraq War were impossible. In a few years we were proved wrong (with a capital W). Based on this experience, I warn you that my credentials as a predictor of the future are somewhat suspect.

The authors of this book are hopefully better at prognostication than me and my colleagues proved to be. They have done a workmanlike job of organizing the trends and developments of the present into an excellent framework for viewing the vast array of future possibilities. Here you will find short analyses of clusters of important ideas and trends centered on many central themes. It is inevitable that some of these ideas will come to dominate the shape of our future world in the coming decades while others will ultimately fall away as quaint but interesting notions that did not pan out. You will be able to consider this large range of possibilities. And having done so, you will be better prepared for what is to come.

You can also choose for yourself which of the trends that are outlined here will prove to be of lasting importance. If I were pressed to choose the themes that would ultimately have the greatest impact, I would select quantum entanglement and computing, life extension and eliminating aging, asteroid mining, and artificial intelligence. However, as I have indicated above, my track record as a prognosticator is somewhat suspect.

John G. Cramer
Emeritus Professor of Physics
University of Washington
Seattle, Washington, USA
May 29, 2019

The Elegant Future

Introduction

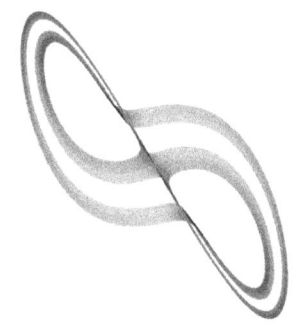

This is a book about ideas. It is about ideas concerning the future and how amazing this world is. It is a book about optimism in a world where it is in short supply. The writings in this book come from long phone conversations between the authors that occurred over the last forty-plus years that were about optimism, the future, and the advancement of technology and knowledge. This book is an attempt to bottle some of those conversations and put a cork in it, so that others make take a swig and hopefully gain some benefit from the experience.

This book was originally supposed to have three authors, but Edward Nixon died, and his sudden departure from this Earthly realm gave the book a sudden sense of urgency. Time is growing shorter for us, and the amount of years behind us is now larger than the predicted years ahead. Too much time is being spent by others on silly politics and chasing after old stale ideas that are doomed to fail. There is so much that needs to be done in this world, but the mainstream media doesn't seem to care about it. Movies and television have become overwhelmingly dystopian in nature, showing dark, bleak futures full of struggle and misery. The future should be one filled with promise, joy, and excitement instead. There is so much more in this world to explore, discover, and learn. Yet many people continue to bounce around in tiny bubbles of perception, creating their own prisons of limits and hopelessness. They seem unaware of how vast and wonderful this reality is, and how unlimited its resources and potentials actually are.

This book is a collection of essays loosely organized by their content. They are each fairly self-sufficient, so they can be read in any order. The book can be read from front to back, or back to front. You may choose to scan or skip topics that are not of interest. Whatever method you use to read this book, we hope you enjoy it as much as we enjoyed writing it.

Craig Philip Peterson

Prologue: Toward A More Elegant Future

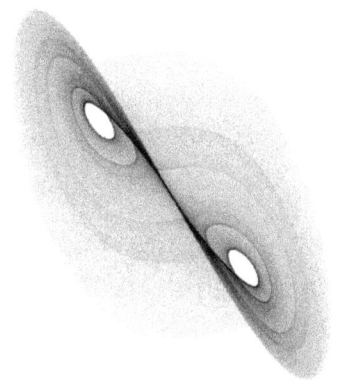

This book is intended to be revolutionary. The authors believe the human race deserves a better future that what we see in entertainment, and in what is discussed in public dialog about the future. We believe that we need an Elegant Future. This book will begin the dialog intended to create such an Elegant Future, by at first describing briefly what such a future might entail and then how we get there and what we might find once we live in An Elegant Future.

AN ELEGANT FUTURE

The Elegant Future

In such a future the human race will be out in the solar system, moving in vast luxury-liners across the planets, moons, asteroids, and artificial habitats of our interplanetary neighborhood. We will develop technology that will allow us to operate in a shirt-sleeve environment on the surface of many of those places.

Every human will have an economic stake in the future via an ownership society of the asteroids. There will still be income disparity, but it will be between the merely wealthy and the super-wealthy. Today's Culture of Cruelty will be greatly lessened and there will be wealth, land, and opportunity for everyone.

We will be on the cusp of developing technologies that will radically transform human life. Imagine the new technologies underway in such An Elegant Future: Instant Universal Communication (better than Star Trek's subspace radio); Profuse Field Teleportation (better than Star Trek's transporters); and Jump Drives (derived from scientific knowledge of the basic underlying structure of sub-atomic particle movement in the universe, much faster than Star Trek's Warp Drive). Those developments and many more will be covered in this book.

HOW DO WE GET THERE?

One basic maxim should guide our thoughts and efforts: CULTURE FLOWS TOWARD POLITICS.

Storytelling, entertainment at its most basic, both gives us ideas about the future and expectations about that future.

Ideas About the Future – EXAMPLE: A Motorola engineer is watching Star Trek, sees Captain Kirk take out his communicator, and thinks to himself (paraphrasing): *"I can make one of those."* He does. We call it the cell phone, or smart phone, and today's smart phones do a lot more than Kirk's communicator ever did.

Expectations About the Future – EXAMPLE: The Jetsons, a mid-20th Century animated series popularized the concept of aircars (more precisely personalized airborne transport) to the point that many of its viewers expected them by the early 21st Century, and disenchantment with today's future often begins with: *"where are the Jetson's aircars?"*

So, if we want, expect, and demand a better, more **Elegant Future**, it must begin with our expectations, which are not being met in today's visual Science Fiction.

The lights go down. You sit in the theater, hunched down in a seat, waiting past the adds. Yes, you've silenced your cell phone. You're waiting for an exciting tale set in a fabulous background in an exciting future. Oh, sure there will be dangers. Or maybe you're watching your television set; you could get stuck on a planet under Klingon occupation. Better hope the locals are actually a highly advanced bunch of energy beings. But despite the dangers, you would like to live in that future.

Or it's decades ago, it's raining outside, the winds skirls a mighty tune, whipping around the house, while in the distance, thunder booms like John Bonham pounding the skins, whilst John Entwistle puts up a furious bass backbeat, Jimmy Page's guitar climbs ramping up in volume and speed, whilst Robert Plant sings of Golem and the Evil One and of Kashmir, as you dive into the latest Asimov foundation story, or a Poul Anderson yarn about Dominic Flandry, or a James Schmitz tale about the fantastic adventures of Telzey Amberdon and her friend Trigger Argee.

No matter what travails await Golan Trevize, what skullduggery Flandry's many enemies are plotting against him, or what monsters stalk the two distaff agents of the Hub, Telzey the teenage telepath and her mission-hardened colleague, Trigger, who's fast with the – well – trigger on her trusty blaster, they inhabit futures that while dangerous are still colorful and exciting – and Elegant.

But more recently as you wait for a series, or movie to be streamed, the stories have grown less colorful, and more predictable. They'll be set in the future, and either their protagonists will be fighting for limited resources in a Mad Max-style low-tech dystopic future, or they'll be battling out in the solar system in a future of scarcity, where the Belters are exploited, and Earth and Mars are fighting it out for solar hegemony. Everything is dirty and dingy, it isn't a future anyone would actually want to live in, and it looks like what the characters probably smell like – bad, certainly a future that's not Elegant.

This is a major problem. Science Fiction should issue warnings, true (it's said a futurist foresees cars, while a science fiction writer sees a traffic jam). But Science Fiction should also point the way to a better future, as shows like Star Trek and writers like Isaac Asimov certainly did. Without an optimistic vision, people's dreams will perish.

What vision do we want for the future? What makes a future an Elegant Future? In the next several chapters we will look at how society has to evolve to create an Elegant Future.

The initial question is why has this tendency to write dystopias happened? This can be answered simply: when technological innovation outpaces societal and political capacity to adapt to meet changing conditions, then we have social chaos. And social chaos can lead to a collapse of a society, especially one as vulnerable to the rapid movement of perishable goods such as our own.

What can be done to combat this growing consensus of future disaster (in addition to insisting on more portrayals of positive futures in fiction)?

We can insist on better, more enlightened public policies in tune with a rapidly changing technological society. In the recent past we saw the awakening of the public to some of the conditions preventing us from growing toward an Elegant Future. And one potential paradigm shifting protest.

OCCUPY WALL STREET: The grassroots protest that worked

The largest and most important manifestation of this understanding was the true grassroots protest against an economy that is producing more have-nots than haves and increasing the disparity between the two. The protests were called collectively, albeit somewhat clumsily and inaccurately: Occupy Wall Street.

Chaotic and unfocused the protests had no apparent leadership or long-term goals and/or policy proposals, but – maybe because they were truly a grassroots unplanned protest – they spread and for a while every large and some medium size cities had protesters camped out on public spaces.

Many spoke out against unfair economic and legal policies that prevented the poor from upward mobility. (See Chapter End Note)

Beyond merely protests we need programmatic action in what we term Three Necessary Steps:

First -- end the Culture of Cruelty.

Second -- embrace the concept of an Ownership Society.

Third -- develop advanced transformative technologies that will make us wealthier, freer, and more emotionally fulfilled.

We will expand upon these three areas in chapters in the book.

END NOTE:

One of the authors of this book (Mr. Elliott) was asked to speak at one of the rallies, and he argued in his remarks that we needed to change securities laws to allow unaccredited investors (those with low incomes – the laws changed the definition of the aforesaid over time) to invest in start-ups.

Thanks to securities law activists (including at least one former Securities and Exchange Commission litigator) eventually Congress changed the rules (realizing that the rise of the Internet acted as a democratizing leveler of information) thus allowing the middle class and the poor to invest in start-ups under certain conditions (which vary greatly state-by-state).

While the current system is not perfect, it does allow for vastly more upward mobility. By that standard the Occupy Wall Street Movement was an unqualified success.

The Elegant Future

Elton Elliott

PART 1:

THE AMAZING WORLD WE LIVE IN

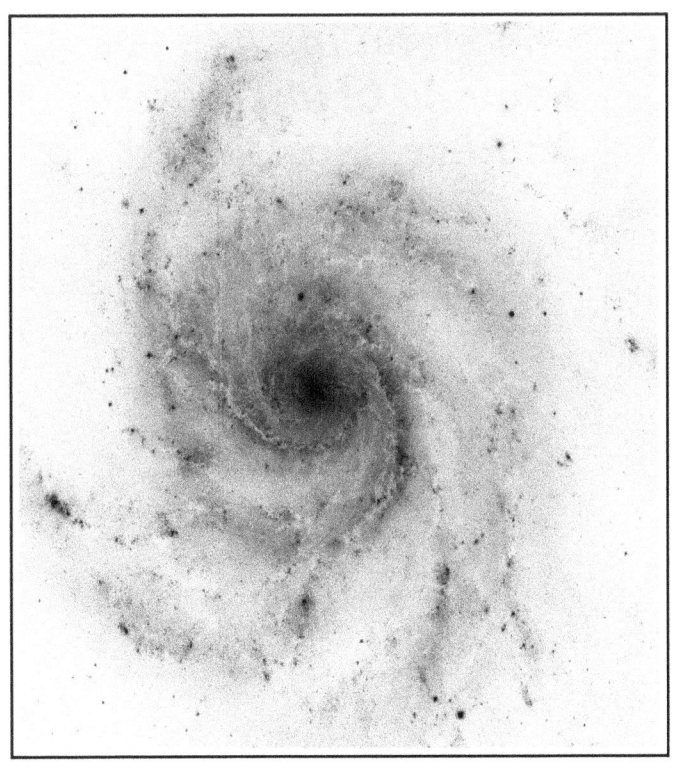

The most beautiful thing we can experience is the mysterious. It is the source of all true art and science.

--Albert Einstein

Man's Place in the Universe: The Unlimited Frontier

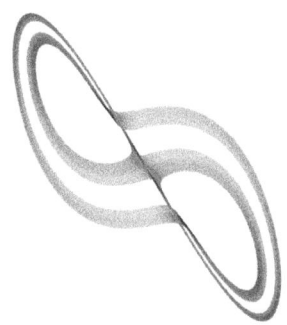

     The Earth is a ball of hot metal and rock 12,742 kilometers in diameter (3,958.8 miles) with a solid metal core, a mantle made of hot, mostly solid, semi-plastic rock, and a thin rocky crust, 71% of which is covered with a thin layer of water. The Earth rotates around a G-type main-sequence star (a very common type of star present in the Milky Way Galaxy) which we call the Sun. The Earth rotates around the Sun every 365.26 days at an average distance of 149.6 million km (or 93 million miles), while itself rotating on an axis with a 23.4 degree tilt every 23 hours and fifty-six minutes.

     There are several extraordinary facts about the Earth that make it a perfect habitat for the human race and the myriad other forms of life that exist on it. The first is its size. If the Earth were much larger, the gravitational pull would make it difficult for upright beings like man to move about, and we would have been forced to crawl about on our bellies. It's far harder to develop a civilization when you're down on all fours. If it were smaller and had less gravity, the Earth would have gradually lost its protective atmosphere leaving the planet a barren rock. The Earth's spinning molten metal interior is currently generating a large magnetic field that protects the life on this planet from the Sun's deadly radiation. Without the magnetic field's protection, the protective ozone layer would be blasted away, and life on Earth would be sterilized by ultraviolet radiation. Another factor is the Moon. Its proximity to the Earth drives the ocean tides and helps to shape the planet's weather. It also stabilizes the Earth's rotation allowing for predictable seasons. Tidal heat transfer may also help stabilize the planet's climate. The Earth's oceans also have a mitigating effect on the planet's climate, absorbing heat and releasing it gradually for more uniform global temperatures. The Earth is also in what astronomers call the 'Goldilocks zone,' which is the region of space surrounding a star which is believed to be the correct distance for life to develop. Having any of these factors absent would have had a dire consequence for the future of the Earth's biology.

Orbiting the Earth is the Moon. The Earth's moon is the fifth largest moon in the Solar System at 3,474.2 kilometers in diameter (or 2,158 miles). It orbits the Earth every 27 days at 384,400 kilometers distance (or 238,900 miles). The large size of the Moon makes it fall into the category of being a minor planet, which makes some groups consider the Earth-Moon system a double planet. Having essentially another planet so close by was a big factor in luring mankind into space. The Moon has always been an object of fascination and folklore. It was manifest destiny for man to travel there as soon as it was technologically feasible.

The Moon also has many extraordinary characteristics that seem to go against random probability. One such trait is its large size compared to the Earth. The larger size allows for a reasonably strong gravitational field, but not so powerful as to make manned landings expensive. The second advantage of the Moon is the tidal locking that keeps one face always toward the Earth. Once a communications antenna is correctly pointed at the Earth, it won't require any further adjustments. Scientists recently found water ice on the Moon, a vital resource for a manned lunar presence. In addition, the

Moon is lined with lava tubes, which are potentially the perfect place to construct large habitats beneath the surface and out of reach of deadly solar radiation. The scale of these lava tubes cannot be overexpressed. Some of them are many kilometers in length with ceilings that are hundreds of meters high, capable of housing major cities with room to spare. These underground caverns are the most likely places where water deposits can be found as well, a bonus for a manned base. These characteristics make the Moon a perfect jump off point for exploring the rest of the known universe.

THE ENTRANCE TO A LUNAR LAVA TUBE LOCATETD IN THE SEA OF TRANQUILITY

The Sun is at the center of the solar system. It is a massive source for energy, radiating 3.846×10^{26} watts per day. When compared to man's meager energy production, this is an impossibly large amount of power (see chapter on fusion power). It is more than enough energy to support any possible requirement. Just out from the Sun, the planets Mercury and Venus are far too hot to make suitable habitats, at over 800 degrees Fahrenheit, their surfaces are nearly as hot as a blast furnace. Traveling in the other direction things get more interesting. The planet Mars, although smaller and colder than Earth, has days that are nearly the same length as days on Earth (24 hours and 37 minutes) giving it a familiarity that would help mitigate the strangeness of living on another world. Mars also has vast quantities of water and potentially has bacterial life. The landscapes of Mars, although barren, bear strong resemblances to certain desert regions on the Earth. Living on Mars would be a challenge, but completely possible with the help of technology.

Beyond Mars is something even more attractive, the Asteroid Belt.

This vast collection of orbiting debris is a virtual cornucopia of mineralogical wealth of almost unimaginable proportions (see The Economics of Asteroid Mining). Beyond that are the four gas giants Jupiter, Saturn, Uranus, and Neptune, which contain vast reserves of hydrogen gas, which can serve as both rocket fuel, and a source of deuterium for fusion reactors (see the Chapter on Fusion Power). Another function of these giant planets is to act as deflectors for objects entering the solar system, essentially protecting the inner planets. This functionality was actually observed for the case of Jupiter when comet Shoemaker-Levy was photographed crashing into the atmosphere of the gas giant.

Comet ShoeMaker-Levy hitting Jupiter

1024x1024 Near-Infrared Camera
University of Hawaii 2.2-meter telescope

Beyond the gas giants lies the Kuiper Belt. It is similar to the Asteroid Belt, but colder and nearly 200 times larger. The Kuiper belt contains mostly icy objects, unlike the rocky and metallic objects in the Asteroid Belt. The Kuiper belt exists from 30 to 50 AU's out from the Sun (one AU or Astronomical Unit is about 150 million kilometers, or approximately the distance between the Earth and the Sun). It is yet another source of raw material at the far edge of the solar system. The dwarf planet Pluto is the largest object in the Kuiper Belt.

Beyond the Kuiper belt lies the Oort Cloud. The Oort Cloud is a gigantic ring of comets and other objects located between 1000 AUs to 100,000 AUs out from the Sun. Because of its great distance from the Sun, little is known about the objects in the Oort Cloud except when one comes swinging in on a highly eccentric elliptical orbit such as a comet. There are potentially other planets in this vast cloud, but the faint light from the Sun and the great distance makes everything there hard to spot.

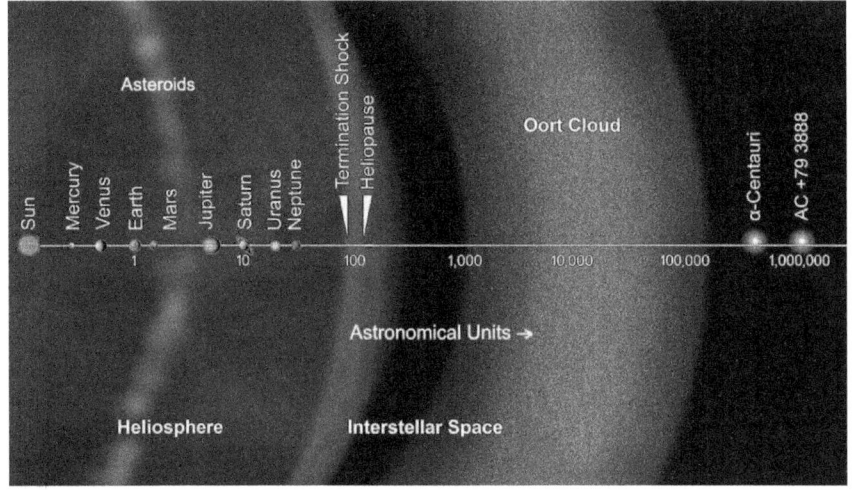

Map of the Solar System

Beyond the Oort cloud lies the vast emptiness of interstellar space. The nearest star system to the Sun is a triple star system known as the Alpha Centauri / Rigel Kentaurus multiple star solar system. The closest of the three stars is Proxima Centauri (or Alpha Centauri C), which is a red dwarf star. Proxima Centauri lies approximately four light years away from the Sun (it would take a beam of light four years to reach it). The other two stars are called Alpha Centauri A (also known as Rigel Kentaurus) and Alpha Centauri B. Alpha Centauri A is a Class G star like the Sun, but slightly larger. Alpha Centauri B is a Class K star, which means it is slightly smaller than the Sun. Both stars are about 4.3 light years away. Using present technology, it would take us 54,000 years to reach this system (based on the velocity of the New Horizons probe, which currently is the fastest man-made object).

The next closest star to the Sun is Barnard's Star. Barnard's Star is a red dwarf star located about 6 light years away. *This star is of particular importance to me because of a dream I had long ago concerning this star. In the dream, a probe was sent to Barnard's Star to an Earth-like planet that was orbiting it. The probe found a moon orbiting the planet which was covered with a gigantic frieze depicting humanoid beings in torment. Further study by the probe of the Earth-like planet indicated there had been a civilization there that had been utterly destroyed. Orbiting around the moon the probe discovered what appeared to be ether a gigantic creature or organic machine. Its appearance was reminiscent of the artwork of the late H. R. Giger. The odd thing about this dream is that the co-author of this book had the **exact same dream**.* Now it turns out that an Earth-like planet *has* recently been discovered orbiting Barnard's Star. If this planet has a moon I will start to worry.

Moving further outward is the rest of the Milky Way Galaxy. The Milky Way is a barred-spiral galaxy, which means it has the general shape of water running down a drain. It turns out this is an accurate description, for there is a gigantic black hole (or a collapsed star that has an escape velocity greater than the speed of light) at the center of the galaxy pulling all the stars toward it and slowly consuming them. Scientists have determined it is highly probable that all galaxies have such hungry monsters at their centers.

The Milky Way is approximately 100,000 lightyears across and contains between 200 billion and 400 billion stars. At the middle of the Milky Way Galaxy is a structure dubbed the 'Galactic Bar.' This is not an alien pub, but rather an elongated region of closely packed stars surrounding the central black hole. Two main arms of stars spiral outward from the ends of the galactic bar. They are called the Scrutum-Centaurus Arm, and the Orion-Cygnus Arm. Our sun is located in the Orion Cygnus Arm. Other smaller arms of stars branch off from these two main arms. The other arms are the Cantina-Sagittarius Arm, the Perseus Arm, the Norma Arm, the Outer Arm, and the New Outer Arm.

The Elegant Future

MILKY WAY GALAXY

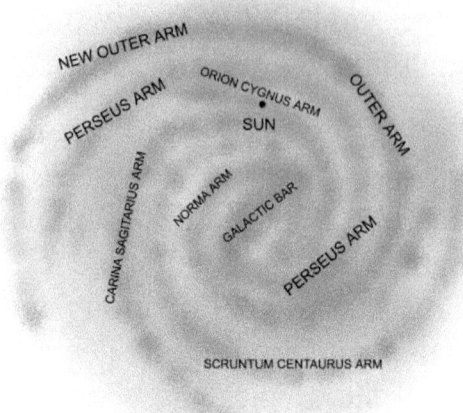

    The stars in the Milky Way move around its gravitational center causing the galaxy to slowly turn like a pinwheel. This is a bit of a problem because the Laws of Physics (namely Newton's Law of Gravitational Attraction) predict that the shape of the galaxy should be an ungodly wound up mess, not a beautiful spiral. Stars farther from the center should orbit more slowly than stars closer to the middle, but this is not entirely the case. In fact, the stars in the outer parts of the spiral arms are moving faster than they should be. This contradiction is called the Galactic Winding Problem. One explanation for this discrepancy is there is more invisible mass in the galaxy than we are able to detect. This extra galactic mass fudge factor has been dubbed dark matter. For the dark matter to have the proper correcting effect, over 78% of all solid matter must be Dark Matter.

There are many smaller minor galaxies orbiting around the Milky Way Galaxy. The current count of satellite galaxies is 59, and that number keeps getting higher. The ten largest satellite galaxies (from largest to smallest) of the Milky Way are:

1. The Large Magellanic Cloud Galaxy
2. The Antlia 2 Galaxy
3. The Sagittarius Dwarf Galaxy
4. The Crater II Galaxy
5. The Small Magellanic Cloud Galaxy
6. The Canes Venatici I Galaxy
7. The Canis Major Dwarf Galaxy
8. The Bootes III Galaxy
9. The Sculptor Dwarf Galaxy
10. The Draco Dwarf Galaxy

All of these Galaxies lie between 8 and 250 kiloparsecs from the Milky Way Galaxy (1 kiloparsec = 3.262 light years). They range in size from about 13,000 light years across to less than 100 light years in diameter. The Milky Way and the Large Magellanic Cloud are predicted to collide in about 2.4 billion years.

The next major galaxy is the giant spiral galaxy known as the Andromeda Galaxy. The Andromeda Galaxy has more than twice the number of stars and is over twice the size of the Milky Way Galaxy with a diameter of more than 220,000 light years. It is on a collision course with the Milky Way Galaxy, and that catastrophe is expected to occur in about 4.2 billion years. The Andromeda Galaxy is the largest member of a group of nearby galaxies known as the Local Group.

The Triangulum Galaxy is the next closest galaxy in the Local Group. It is a small spiral galaxy that measures about 60,000 light years across. Triangulum is currently on a hyperbolic orbit around the Andromeda Galaxy. It is not clear whether the two will eventually collide. Another interesting fact about the Triangulum Galaxy is that it is currently emitting two gigantic water masers from opposite sides of its spiral form.

The Local Group of galaxies is a clump of more than fifty galaxies nearest our own galaxy. They comprise a region of space that measures about 10 million light years across. These galaxies are part of a larger group of galaxies known as the Virgo Supercluster. The Virgo Supercluster is part of an even larger strand of galactic clusters known as the Pisces–Cetus Supercluster Complex. This supercluster strand is part of an even greater cosmic watershed known as Laniakea.

Laniakea is called a cosmic watershed because it was discovered by determining the movements of the galaxies within its boundaries. A large number of the galaxies in Laniakea are all drifting toward a single point which has been dubbed the great attractor. Other parts of Laniakea include such mysterious sounding entities as the Coma Cluster, the Leo Cluster, the Shapley Cluster, and the Perseus-Pisces cluster. Laniakea resembles the Norse cosmological tree known as Yggdrasil which joined various mythical worlds together.

Laniakea is just a small part of an unimaginably large web-like structure in the greater universe. The number of galaxies estimated in the entire universe is as high as two trillion, each containing hundreds of thousands of stars. This large cosmic web bears an uncanny resemblance to neurons in a gigantic brain.

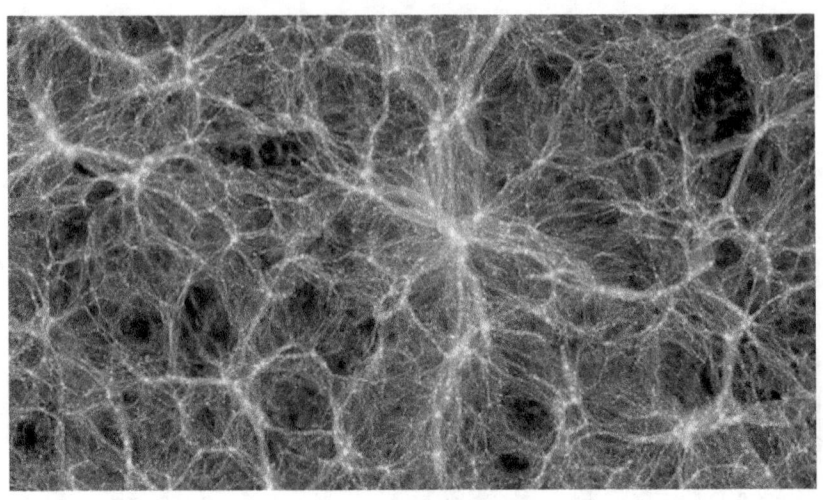

The large-scale web-like structure of the universe.

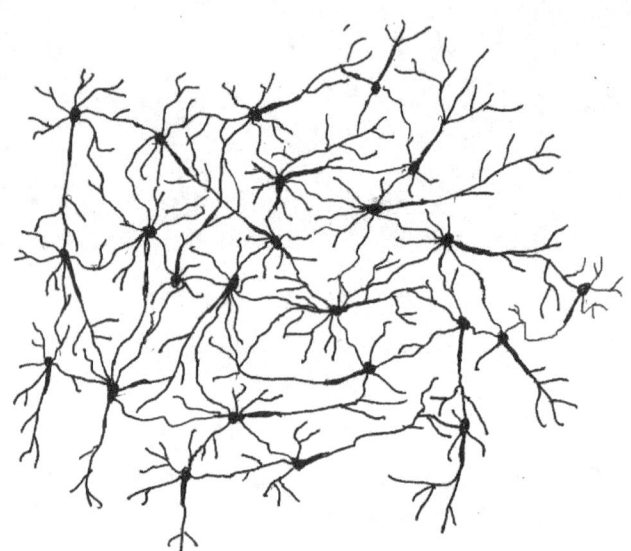

INTERCONNECTING NEURONS

There are more stars in the universe by far than all the grains of sand in all of the beaches and all of the deserts in the world combined. The universe is really big and getting bigger every day. If fact, on the average other galaxies are not only moving apart from each other, but they are accelerating away from each other. This behavior violates our current understanding of physics, so another fudge factor had to be introduced called dark energy. Dark energy is energy we are unable to detect. In order for the dark energy correction to work, 73% of the mass of the universe must be composed of dark energy.

The size of this universe is so large that it is beyond human capacity to fully comprehend it, and it appears to be getting bigger at an accelerating rate. This vastness is only what we are able to observe. Our view of the universe is limited to how far light can travel in the time since the big bang occurred. Nobody truly knows what other wonders lurk beyond the range of humanity's crude instrumentation. One fact is certain, with all of this vastness, there is more than enough space for humanity and whatever else is out there.

The Elegant Future

Reality – What a Concept

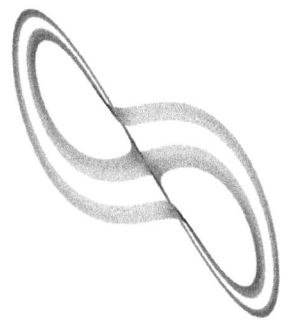

What is reality? Will our view of reality in the distant future be the same as it is now? Can reality be measured only by space and time, or are there other aspects to reality that we must eventually consider? Is reality limited to what our senses convey, or are we only seeing a tiny fraction of the truth of it? How big is reality and what is its shape? Is it of fixed size or is it limitless? If you travel far enough will you disappear into the void, or will you end up where you started? What about the resolution of reality? Is there a limit to how small you can get? If there is a limit, what does the smallest piece of reality consist of? Is empty space really empty? Are there other realities than the one we currently exist in? If so, will we ever be able to travel there? Do others perceive the same reality that we do, or do strange discrepancies exist? The questions go on and on.

The Elegant Future

As long as these questions have been pondered, the human race has proposed answers to them. Many of the theories about reality have been outlandish to the extreme. Hindu cosmologists claimed that the Earth sat upon the backs of sixteen elephants. There were eight male elephants and eight female elephants. The names of the male elephants were Airavata, Pundarika, Vamana, Kumunda, Anjana, Pushpa-danta, Sarva-bhauma, and Supratika. They didn't bother to name the female elephants. The elephants in turn stood on the shell of a giant turtle named either Chukwa, Kurma or Kurmaraja. The turtle swam in a great infinite ocean.

The ancient Egyptians believed that reality was an infinite expanse of water that was also a god named Nun. The earth was another god named Geb. The sky goddess Nut arches over the earth. The sun god Ra travels over the body of the sky goddess. Finally, the air goddess Shu was positioned between Geb and Nun.

According to Norse mythology, reality is composed of nine distinct realms called Niflheim, Muspelheim, Asgard, Midgard, Jotunheim, Vanaheim, Alfheim, Svartalfheim, and Helheim. These realms are nestled in the branches and roots of the world tree Yggdrasil. Asgard sat in the top branches of the tree. Helheim was located beneath the roots. Midgard, the earthly realm, sat on the central branches. The other realms were scattered between the branches and roots. The interesting aspect of the Norse reality is that recent scientific discoveries have shown that this idea wasn't completely wrong (see Laniakea in Man's Place in the Universe).

The Judeo-Christian view of reality (although still wrong) was far more practical in its construction. There were no turtles or elephants or multiple gods or giant trees. The earth was essentially a large island resting on pillars in a vast primordial sea. At the edges of this earth-island sat the pillars of heaven which held up a giant dome called the firmament. The stars were tiny holes in the firmament through which the light from heaven shined. The Moon and Sun moved around on the underside of the firmament. Beneath the surface of the earth was a great hot cavern that was the underworld (or hell) where damned souls were tortured indefinitely. Above the firmament was the ocean of heaven (the source of rain). Above the ocean of heaven was the heaven of heavens.

The Elegant Future

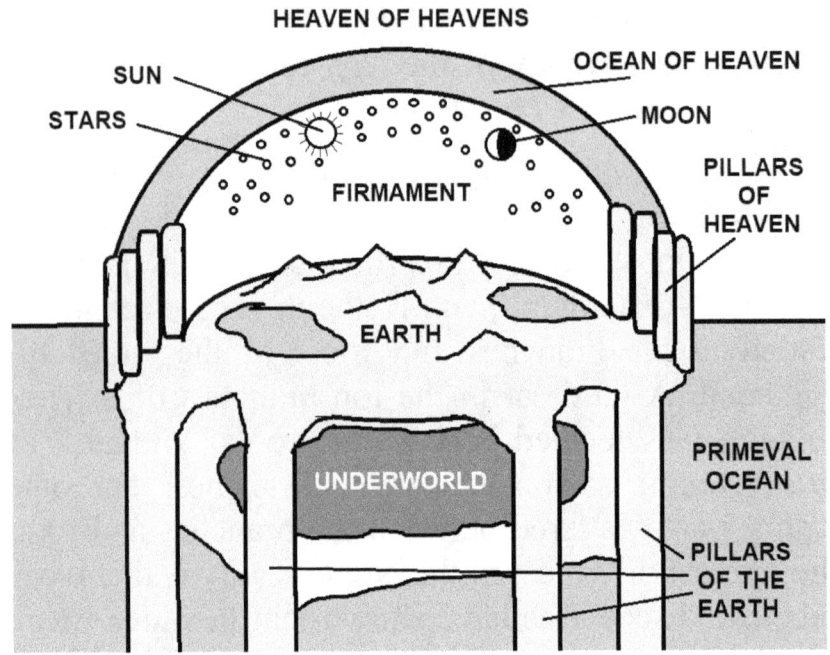

Eventually scientific observations, measurements, and computations caused the early outlooks of reality to be gradually modified until our present perspective came into fruition. The Earth is not at the center of all things, the universe is far larger than we can even comprehend, and there are things lurking in the vastness that make the old tales of hell seem tame in comparison (namely magnetars and super giant black holes). Our instruments for perceiving reality are ever growing in capability. It is of little doubt that aspects of what we believe about reality today will require further modification in the future.

EARTH   AIR   FIRE   WATER

Putting aside the changes in the macroscopic view of reality. The microscopic view has also changed. The idea that all material is composed of constituent elements is an ancient one. The ancient Greeks, Babylonians, Japanese, Indians, Egyptians, and Tibetans all believed these elements were Earth, Air, Fire, and Water (occasionally the Void was added as a fifth element). As this was later disproved (partly by alchemists trying to transmute lead into gold and failing), there was a gradual realization that reality was more complex than the experts had believed. In 1789, Antoine Lavoisier published *Elements of Chemistry,* which was the first modern list of chemical elements. The list contained thirty-three elements, including light and caloric. Lavoisier was later beheaded during the French Revolution partly because *"the Republic has no need of scientists or chemists."* Eventually the modern periodic table was constructed and largely completed by 1914.

Periodic Table of the Elements

Legend: atomic number / symbol / atomic weight (e.g., 1 H 1.008)

1	2	3	4	5	6	7	8	9	10	11	12	13	14	15	16	17	18
1 H 1.008																	2 He 4.003
3 Li 6.941	4 Be 9.012											5 B 10.81	6 C 12.01	7 N 14.01	8 O 16.00	9 F 19.00	10 Ne 20.18
11 Na 22.99	12 Mg 24.31											13 Al 26.98	14 Si 28.09	15 P 30.97	16 S 32.07	17 Cl 35.45	18 Ar 39.95
19 K 39.10	20 Ca 40.08	21 Sc 44.96	22 Ti 47.88	23 V 50.94	24 Cr 52.00	25 Mn 54.94	26 Fe 55.85	27 Co 58.93	28 Ni 58.69	29 Cu 63.55	30 Zn 65.39	31 Ga 69.72	32 Ge 72.59	33 As 74.92	34 Se 78.96	35 Br 79.90	36 Kr 83.80
37 Rb 85.47	38 Sr 87.62	39 Y 88.91	40 Zr 91.22	41 Nb 92.91	42 Mo 95.94	43 Tc (98)	44 Ru 101.1	45 Rh 102.9	46 Pd 106.4	47 Ag 107.9	48 Cd 112.4	49 In 114.8	50 Sn 118.7	51 Sb 121.8	52 Te 127.6	53 I 126.9	54 Xe 131.3
55 Cs 132.9	56 Ba 137.3	57–71*	72 Hf 178.5	73 Ta 180.9	74 W 183.9	75 Re 186.2	76 Os 190.2	77 Ir 192.2	78 Pt 195.1	79 Au 197.0	80 Hg 200.6	81 Tl 204.4	82 Pb 207.2	83 Bi 209.0	84 Po (210)	85 At (210)	86 Rn (222)
87 Fr (223)	88 Ra (226)	89–103‡	104 Rf (261)	105 Db (262)	106 Sg (263)	107 Bh (262)	108 Hs (265)	109 Mt (266)	110 Ds (271)								

Numbers in parentheses are atomic mass numbers of radioactive isotopes.

*lanthanide series

57 La 138.9	58 Ce 140.1	59 Pr 140.9	60 Nd 144.2	61 Pm (145)	62 Sm 150.4	63 Eu 152.0	64 Gd 157.3	65 Tb 158.9	66 Dy 162.5	67 Ho 164.9	68 Er 167.3	69 Tm 168.9	70 Yb 173.0	71 Lu 175.0

‡actinide series

89 Ac (227)	90 Th 232.0	91 Pa 231.0	92 U 238.0	93 Np (237)	94 Pu (244)	95 Am (243)	96 Cm (247)	97 Bk (247)	98 Cf (251)	99 Es (252)	100 Fm (257)	101 Md (258)	102 No (259)	103 Lr (260)

How the laws governing the operating principles of reality were discovered is another example of how man's perception of reality was forced to change as more facts were uncovered. The first real attempt to define natural laws was done by Aristotle. He called these rules the Four Causes. They were as follows:

1. The Material Cause (the composition of matter)
2. The Formal Cause (the arrangement of matter)
3. The Efficient Cause (the causality and agents of change)
4. The Final Cause (the purpose or function)

The four causes were an interesting outlook on the acting agents in reality, but not a lot of help if you're trying to build a bridge (or a better mouse trap).

In 1667, Sir Isaac Newton published *Philosophiæ Naturalis Principia Mathematica* ("Mathematical Principles of Natural Philosophy"). In this one work the modern science of Physics was born. It was this work that contained the laws governing the behavior of matter (under classical conditions). These are Newtons three laws of motion. These laws can be stated as follows:

1. A body will remain at rest or move at a constant velocity unless it is acted on by an unbalanced force.
2. The force exerted on a body is proportional to the mass of the body times the acceleration of that body.
3. If two bodies exert a force on one another, then the forces are equal in magnitude, but opposite in direction.

Written mathematically:

Newton's Laws

1. If $\vec{v}_m = C$ or $\vec{v}_m = 0$

   Then $\sum_j \vec{F}_{m_j} = 0$

2. $\vec{F}_m = m\vec{a}$

3. $\vec{F}_{12} = -\vec{F}_{21}$

Newton's laws are the backbone of modern mechanical and structural engineering. They are just as valid today as when Sir Isaac wrote them down over 350 years ago. The other physical law contained in Principia Mathematica is Newton's Law of gravitation, which states that *the gravitational force between two masses is proportional to the product of the two masses divided by the square of the distance between the centers of the masses.*

This is expressed mathematically as:

$$F_G = G \, \frac{m_1 m_2}{r^2}$$

F_G = Gravitational Attraction Force

G = Gravitational Constant

r = Distance Between the Centers of the Two Masses

m_1, m_2 = The Two Masses

This seventeenth century formula is still used to calculate the trajectories of spacecraft today.

The classical formulas of Newton described the behavior of masses and gravity, but what about the mysterious forces of electricity and magnetism? The first descriptions for understanding electro-magnetic reality was defined by French physicist Charles Augustin de Coulomb (1736–1806). He determined that the electric force behaved similar to the gravitational force in that *the force between two charges was proportional to the product of the charges divided by the square of the distance between them.* This is written mathematically as:

$$F_e = k_e \frac{q_1 q_2}{r^2}$$

F_e = Electric Force

k_e = Coulomb's Constant = $\frac{1}{4\pi\varepsilon_0}$

r = Distance Between the Centers of the Two Charges

$q_1 q_2$ = The Two Charges

ε_0 = The Permittivity of Free Space

This equation is nearly identical to Newton's Law of Gravity, the main difference being that since charges can be positive or negative, the force between charges can be either attractive or repulsive.

Another major contributor to our understanding of the electromagnetic phenomena was German mathematician Carl Fredrich Gauss (1777–1855).

Gauss formulated Gauss's Law which states that *the net electric flux through any closed surface is proportional to the net electric charge within that closed surface*. This is written mathematically as:

Gauss's Law

$$\nabla \cdot E = \frac{\rho}{\varepsilon_0}$$

$$\nabla \cdot E =$$ The Divergence of the Electric Field

$$\rho =$$ Charge Density

$$\varepsilon_0 =$$ The Electric Constant (permittivity or dielectric constant of free space)

The third major contributor to our understanding of the electromagnetic phenomenon was English physicist Michael Faraday (1791-1867). Faraday discovered the relationship between the electric force and the magnetic force. Faraday's Law states that *the electromotive force (Voltage) around a closed path is equal to the negative of the time rate of change of the magnetic flux enclosed by the path.* This is written mathematically as:

$$\mathcal{E}_{EMF} = -\frac{\partial \Phi_B}{\partial t}$$

$$\mathcal{E}_{EMF} = \text{The Electromotive Force}$$

$$\Phi_B = \text{The Magnetic Flux}$$

It took a Scottish man named James Clerk Maxwell (1831-1879) to assemble these three men's ideas about electromagnetism into a unifying set of equations. They are, of course, called Maxwell's Equations and are written as follows:

$$\nabla \cdot \mathbf{E} = \frac{\rho}{\varepsilon_0}$$

$$\nabla \cdot \mathbf{B} = 0$$

$$\nabla \times \mathbf{E} = -\frac{\partial \mathbf{B}}{\partial t}$$

$$\nabla \times \mathbf{B} = \mu_0 \left(\mathbf{J} + \varepsilon_0 \frac{\partial \mathbf{E}}{\partial t} \right)$$

Where:

$\mathbf{E}$ = The Electric Field

$\mathbf{B}$ = The Magnetic Field

$\mathbf{J}$ = The Electric Current Density

ε_0 = The Permittivity of Free Space

μ_0 = The Permeability of Free Space

(note that $\nabla \times$ is the curl operator)

The top equation is Gauss's Law. The second equation states that the divergence of a magnetic field is zero (there are no magnetic monopoles). The third equation is essentially the vector version of Faraday's Law. Note that the curl operator in the relationship causes the two fields to be perpendicular to each other. The fourth equation is similar to the third in showing the relationship between the fields, but also shows how the changing electric field and the current density is related to the strength of the magnetic field. Maxwell's Equations have revolutionized the human experience. They have allowed men like Nicola Tesla (1873-1943) to develop electric motors and generators. They have led to the creation of electronics and computers. They have forever changed humanity's view of reality.

Another triumph of Maxwell was his assertion that light was actually an electromagnetic wave. Prior theories assumed that light was made of very small particles called photons. This led to a conundrum called the wave–particle duality. In some cases, light acted like particles, in other cases light acted like waves. Some wondered if all matter (not just photons) behaved in a similar way. The answer turned out to be yes. A French physicist named Louis Victor Pierre Raymond de Broglie came up with the equation for matter waves. It is written as follows:

$$\lambda_m = \frac{h}{p} = \frac{h}{mv}$$

$$\lambda_m = \quad \text{The wavelength of a particle of mass m}$$

$$h = \quad \text{The Plank Constant}$$

$$p = mv = \quad \text{The Momentum}$$

$$m = \quad \text{The Mass of the Particle}$$

$$v = \quad \text{The Velocity of the Particle}$$

The idea that all matter in reality was neither particles nor waves, but a weird combination of both led physicists to rethink their perceptions about atoms. The classical view of atoms is that they consisted of electrons orbiting a nucleus of protons and neutrons. The *Electromagnetic Force* kept the electrons in orbit around the nucleus. The protons and neutrons in the nucleus were stuck together by the *Strong Nuclear Force*, and radioactive decay was governed by the *Weak Nuclear Force*.

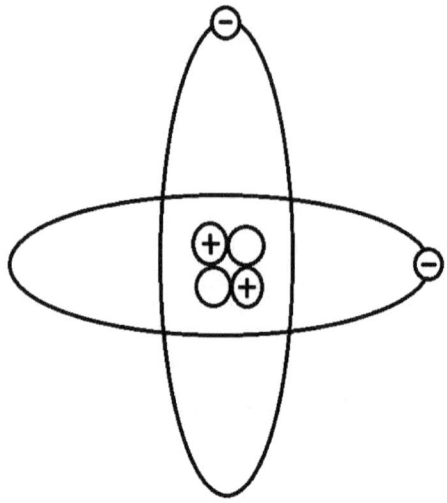

Classical Helium-4 Atom

(Actually the electrons should be sharing an orbit, but this is more asthetically pleasing.)

This view was no longer entirely true if the atom was really wavy electrons jiggling around wavy protons and neutrons. Something better was needed to describe the situation. In 1925 an Austrian physicist named Erwin Rudolf Josef Alexander Schrödinger published an equation to better explain the behavior of matter. It (of course) is known as *the Schrödinger Equation,* and with this equation the field of *quantum mechanics* was born. The Schrödinger equation is written as follows:

$$i\hbar\frac{\partial}{\partial t}\Psi(\mathbf{r},t) = \left[\frac{-\hbar^2}{2m}\nabla^2 + V(\mathbf{r},t)\right]\Psi(\mathbf{r},t)$$

$i = \sqrt{-1}$

$\hbar = h/2\pi$ where h is the Plank Constant

$\dfrac{\partial}{\partial t} =$ The Partial Derivative with Respect to Time

$\Psi =$ The Wave Function

$r =$ The Radius from the Center of the Particle

$t =$ Time

$m =$ The Mass of the Particle

$\nabla =$ The Laplacian Operator (or Divergence)
(the square indicates it is taken twice)

$V =$ The Potential Energy of the Particle

The equation seems complex, but it is really only a new combination of the previous concepts. Quantum mechanics changed atoms from little balls orbiting each other into bizarre clouds of probability distribution. The probability lobes of the clouds were actually the places where the electrons would most *probably* be found. Some electron probability distributions are shown below:

The Elegant Future

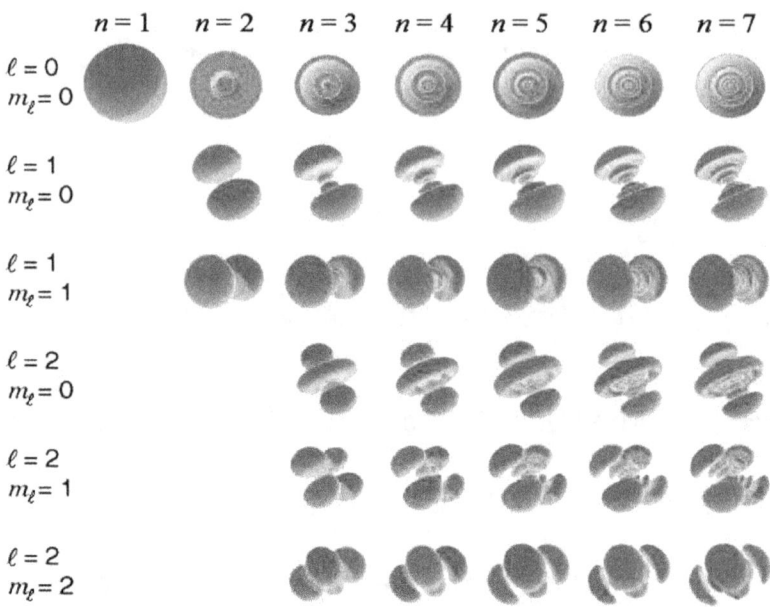

| | $n = 1$ | $n = 2$ | $n = 3$ | $n = 4$ | $n = 5$ | $n = 6$ | $n = 7$ |

$\ell = 0$
$m_\ell = 0$

$\ell = 1$
$m_\ell = 0$

$\ell = 1$
$m_\ell = 1$

$\ell = 2$
$m_\ell = 0$

$\ell = 2$
$m_\ell = 1$

$\ell = 2$
$m_\ell = 2$

$n =$ Electron Energy

$\ell =$ Angular Momentum

$m_\ell =$ Magnetic Quantum Number

The equations governing quantum mechanics have enabled experimenters to make very precise predictions about the behavior of matter. The actual meaning of this somewhat enigmatic theory is open to interpretation. The interpretations of quantum mechanics are many and varied. Some of them are (sorted by date):

1. The De Broglie Bohm Interpretation

2. The Copenhagen Interpretation

3. The Stochastic Interpretation

4. The Many Worlds Interpretation

5. The Consistent Histories Interpretation

6. The Objective Collapse Interpretation

7. The Transactional Interpretation

8. The Relational Interpretation

9. The Ensemble Interpretation

All the various interpretations result from different opinions about the role of the particle's probability wave function and what it actually represents. It's difficult to properly describe these interpretations in a few sentences, but since this is not a book about quantum mechanics, here I go anyway.

The *De Broglie-Bohm Interpretation* insists that the probability distribution is actually a *pilot wave* that the particle is bouncing around in. The theory is named for Louis de Broglie (1892–1987) and David Bohm (1917–1992) and was developed by De Broglie in the 1920s.

The *Copenhagen Interpretation* requires that for the case where multiple outcomes exist, a measurement must be made before a probable outcome

The Elegant Future

can become the actual outcome. This is the famous Schrödinger's Cat thought experiment. (A cat's life depends on a quantum measurement. The cat is neither dead nor alive until the measurement is made.) This interpretation was proposed by Niels Bohr and Werner Heisenberg in Copenhagen in 1927.

The Stochastic Interpretation believes the wavy weirdness is due to spacetime fluctuations. This interpretation was first proposed in 1952 by Hungarian physicist Imre Fényes.

In the *Many Worlds Interpretation*, for every action where there can be multiple outcomes, the universe must divide into a separate universe for each possible outcome (creating a multiverse). The original ideas for this interpretation belong to physicist Hugh Everett in 1957. Later, this interpretation was renamed many-worlds by physicist Bryce Seligman DeWitt in the 1960s.

The *Consistent Histories Interpretation* states that only one outcome is the actual outcome and quantum mechanics just predicts the probability of each outcome. This interpretation was first proposed by Robert B. Griffiths in 1984.

The *Objective Collapse Interpretation* states that the wave function is real, and it collapses to an outcome randomly when some specific physical threshold is arrived at, with the observers playing no special role. The GRW spontaneous collapse theory was published in 1985 by physicists Giancarlo Ghirardi, Alberto Rimini, and Tullio Weber.

The *Transactional Interpretation* is a much more mechanistic view of quantum mechanics. This interpretation was first proposed by Dr. John Cramer in 1986. This interpretation draws its inspiration from the Wheeler–Feynman time-symmetric theory, named after physicists John Archibald Wheeler and Richard Feynman, which contends that the solutions of the electromagnetic field equations must be invariant under time-reversal. In this interpretation, to achieve an outcome the particle has two associated wave functions. One wave function, the retarded wave, travels forward in time, and the other, the advanced wave, travels backwards in time. This in effect creates a handshake or transaction which determines the outcome. This effectively eliminates the necessity for a multiverse and can explain the Bell Inequality's action at a distance. It also requires information to be transmitted backwards in time, which is highly counter-intuitive.

The *Relational Interpretation* suggests that the outcome of a wave function is based on its *relation* to the observer performing the measurement, so different observers produce different outcomes. This interpretation was proposed in 1994 by Dr. Carlo Rovelli.

According to the *Ensemble Interpretation* the wave function does not apply to an individual system such as a single particle, but rather applies to an *ensemble* (a large number) of systems of particles that have the same initial conditions. This interpretation was proposed by Dr. Lee Smolin in 2011.

The Elegant Future

In 1905 Albert Einstein published a paper entitled *"Zur Elektrodynamik bewegter Körper"* ("On the Electrodynamics of Moving Bodies"). The theories put forward in that publication later became known as Special Relativity. Special Relativity asserts the following ideas:

1. The laws of physics are the same in all inertial frames of reference.
2. The speed of light in free space has the same value c in all inertial frames of reference.
3. Mass and energy are equivalent.
4. Nothing can travel faster than the speed of light.
5. The lengths of objects will appear to be shortened in the direction they are moving with respect to the observer.
6. Time passes at different rates in different inertial frames.
7. The mass of an object increases as its velocity increases.

The third idea concerning mass and energy can be written as the famous equation:

$$E = mc^2$$

$$E = \text{Energy}$$

$$m = \text{Mass}$$

$$c = \text{The Speed of Light}$$

Special relativity has had profound ramifications on our perceptions about reality. Time, mass, and physical dimensions are all dependent on their inertial reference frames. In other words, these formerly absolute conditions are all now *relative* conditions.

Einstein later published general relativity which added other ideas to the mix. Some of the ideas are that clocks run slower in gravity wells, light can be bent by gravity wells, and gravity is actually a deformation in space-time.

Physicists have tried to reconcile all these disparate theories into one grand theory which would essentially be the Holy Grail of physics. This theory would connect the fundamental forces of Gravity, Electroweak (Electromagnetic and Weak Nuclear forces were unified by Sheldon Glashow, Abdus Salam, and Steven Weinberg in 1979), and the Strong Nuclear Force. The theory would link relativity with quantum mechanics. This grand unifying theory would explain reality in its entirety. Thus far they have not been successful. These attempts have interesting names. Some of them are:

1. String/M Theory
2. Quantum Loop Gravity Theory
3. Asymptotically Safe Gravity Theory
4. Causal Dynamic Triangulations Theory
5. E-8 Theory

The most popular theory is String/M Theory which proposes that fundamental particles are not the smallest objects in reality, but that the fundamental entity is a one-dimensional object called a string. It also requires extra dimensions (M-Theory has 11 dimensions), and multiple universes (as many as 10 to the 500th power).

Quantum Loop Gravity requires space itself to be composed of a network of tiny finite loops woven together like fabric. These networks of loops are called spin networks. The loops have a scale of approximately 10 to the −35 meters, and smaller sizes do not exist.

Asymptotically Safe Gravity Theory is a theory involving the quantization of gravity that is difficult to describe in non-mathematical terms. The theory uses a renormalization process to prevent the theory from breaking down at high energy levels unlike previous quantum gravity theories (hence asymptotical safety).

Causal Dynamic Triangulations Theory uses a structure called a simplex, which divides spacetime into tiny higher dimensional triangle-like objects known as a 4-simplex or pentachoron *(that's pentachoron not midi-chorlian)*. Each simplex is geometrically flat, but it can be "glued" together with other simplexes to create curved spacetimes.

E-8 Theory uses a 248-dimensional symmetrical mathematical construct known as E-8 (a Lie field object) to represent fundamental particles and their interactions. Its main proponent is a surfing/skateboarding physicist named Garret Lisi.

If you had trouble with this chapter, fear not. It tries to compress four years of college physics into a single chapter. The math could not be omitted because it actually demonstrates the elegant nature of the discoveries made thus far. And more discoveries are yet to be made. As mankind learns more about reality, more questions arise. Perhaps this seemingly vast complexity is much simpler than it currently appears. In 1940 John Wheeler made a telephone call to Richard Feynman and proposed that all electrons and positrons are actually manifestations of a single particle moving backwards and forwards in time. Perhaps reality is manifested by a single fundamental entity traveling back and forth in time making connections and interactions with itself. Or possibly reality can never be fully comprehended. Will there ever be an end to knowledge? Perhaps, but it is unlikely we are anywhere near it. More amazing discoveries are ahead. Like the older theories that fell to more elegant mathematical descriptions, the days for today's theories may well be numbered. Now some scientists propose that reality is not reality at all, but merely a simulation (see hacking the universe). The oddness of quantum mechanics is written off as action at the limits of the reality program's resolution. The physical behavior of matter would be due to algorithms running in the cosmic computer. This begs the question of who's running the simulation? Perhaps reality is pure thought. If so, what does that say about the power of the human mind? Or are we all just thoughts in a far greater mind?

Nature's Grand Machines

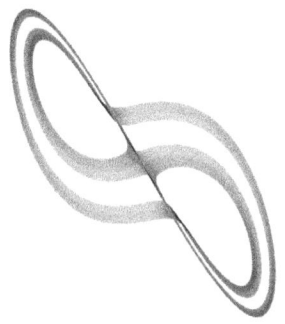

I remember when I first learned about cells. I am not a young person, so this was many decades ago. My mother was the first one to tell me about cells. I was so amazed at the concept, I started looking at every article or book I could find about cells. I was so intrigued by them I started drawing pictures of various cells and labeling them. It became an obsession. My favorite single-celled animal was the paramecium, or slipper animalcule. One day I experienced paramecium firsthand, as they swam about on a microscope slide. It was thrilling to watch them glide around and occasionally bump into each other. Cells were magical things. They were tiny bits of life too small to be seen. They lived on the very edge between life and the lifeless.

Way back then cells were considered to be simple things. They were mere blobs of protoplasm with a nucleus and some other organelles floating around inside their cell membrane. But the idea that these biological units of life were simple was at best a naive notion, at worse vastly ignorant. Cells are not simple at all. They are massively complicated micro machines capable of breathing, eating, self-propulsion, reproduction, and in the case of cells from multicellular organisms, capable of generating a complicated multicellular lifeform of mind-boggling complexity. A human cell can store at least 715 Megabytes of information in its nuclear DNA. This is equivalent to over 7,000 novels. That's enough books to fill a small bookstore. One attempt to model the function of a lowly bacteria required a supercomputer with 42,000 processors.

There are two main types of cells, prokaryotic cells, and eukaryotic cells. Prokaryotic cells are a bit more simplistic than the eukaryotic cell, having no defined nucleus, and a smaller assortment of organelles. The Eukaryotic cell is more complex, has a well-defined nucleus, and a plethora of organelle types. Multicellular animals are almost without exception composed of eukaryotic cells.

The most important organelle in a living cell is its cell membrane. The membrane keeps the outside stuff out and the inside stuff in. At first glance the membrane appears to be very simple – a mere thin film around the cell. Closer inspection reveals that this simplicity is an illusion. A cell membrane is a complex molecular mechanism composed of lipids, molecular transport systems, channel proteins, receptors, complex propulsion systems, and enzymes.

CELL MEMBRANE

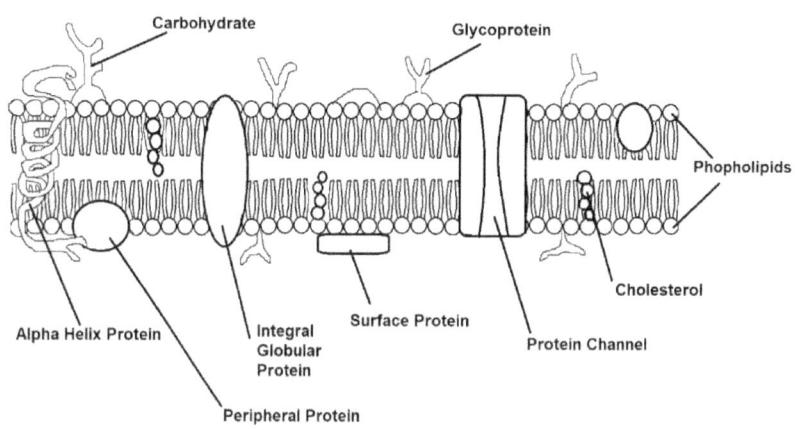

The membrane keeps the cells internal fluid (or cytoplasm) from leaking out. The cytoplasm in a cell is composed mostly of a fluid call cytosol, which is a complex mixture of many different substances dissolved in water. The cytoplasm also contains various organelles, which are essentially the cells organs. The organelles perform a variety of tasks that enable the cell to function properly.

At the center of the eukaryotic cell is the nucleus. The nucleus contains the DNA (Deoxy-ribonucleic Acid) of the cell where all of its operating instructions are located. The nucleus has its own double membrane which is covered with pores to allow mRNA (messenger ribonucleic acid) and ribosomes to pass in and out. It also contains a structure called a nucleolus where ribosomes are created. The nucleus uses the DNA as a template to create RNA which has a variety of functions including coding and synthesizing proteins, folding into enzymes, and regulating various cell functions. Normally the DNA in the nucleus is unwound and floating in a sort of disorganized cloud. When cell division (mitosis) is about to occur, the DNA elegantly winds up into the familiar club-shaped chromosome structures. Each chromosome pair is neatly stuck together at the center with a centromere, which is a specialized sequence of DNA that links a pair of sister chromatids together to form a chromosomal dyad.

THE CELL

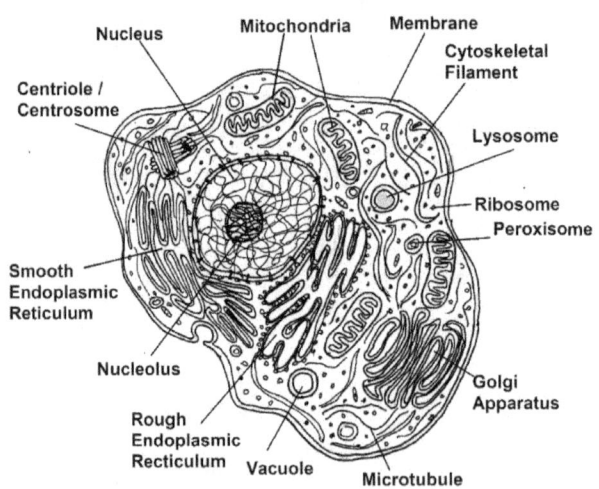

Floating throughout the cell are tiny organelles call ribosomes. Ribosomes use messenger RNA and transfer RNA (tRNA) to assemble amino acids into peptide chains and proteins. (Note: the difference between peptide chains and proteins is simply that proteins are longer.)

Next to the nucleus is a strange convoluted organelle known as the rough endoplasmic reticulum (or rough net-like structure inside the cytoplasm). The roughness is due to the fact it is studded with ribosomes. The rough endoplasmic reticulum's main function is to synthesize and assemble proteins.

Adjacent to the rough endoplasmic reticulum is the smooth endoplasmic reticulum. The smooth endoplasmic reticulum is not covered with ribosomes; hence it is smooth. The smooth endoplasmic reticulum's main function is to manufacture lipids, phospholipids, and steroids.

Another strange organelle is the Golgi apparatus (name for the Italian scientist Camillo Golgi (1843-1926). The Golgi apparatus's purpose is to store proteins or other substances, and then transport them to the Golgi bodies (or Golgi vesicles) for distribution to other locations within the cell.

Mitochondria are another organelle found in cells. These organelles have their own separate DNA which leads scientists to speculate that they evolved from bacteria that lived symbiotically inside the cell, and later became completely dependent on the cells environment to survive. The Mitochondria are literally the batteries of the cell. Mitochondria produce energy by generating adenosine triphosphate (ATP), which serves as fuel for the cell's protein motors. ATP production is part of something known as the Krebs's cycle (also known as the citric acid cycle).

Organelles called lysosomes serve as the cleaning crew for the cell. Lysosomes digest obsolete, un-used, or damaged materials in the cytoplasm.

A very strange organelle pair is the centriole/centrosome pair. This organelle stands out for its very mechanical appearance, looking like two ribbed cylinders connected at right angles to each other. These two organs serve as the main microtubule organizing center for the cell. The microtubules are tiny self-assembling tubes that serve as the cell's main transport system.

Centriole / Centrosome Pair

Vacuoles are organelles that are used to store water or enzymes. They are basically hollow membranes within the cytoplasm.

Peroxisomes are organelles that function to breakdown very long fatty acid chains using a process called beta oxidation. Their structure is similar to vacuoles, but they contain oxidative enzymes (sometimes in crystallized form).

Plant cells contain three additional organelles not found in animal cells. These are the chloroplasts, leucoplasts, and chromoplasts. These organelles are involved in a process known as photosynthesis, in which a plant cell gains energy by absorbing sunlight, carbon dioxide, and water, then converts those input components into oxygen and glucose.

The cell is powered by ATP which acts as the fuel for a variety of small chemical motors made out of protein. These motors are very mechanical in their nature, complete with interconnecting parts, joints, and even rotating wheels. The most bizarre protein motors are the walking protein motors. These protein walkers are known as the kinesin walker, and the dynein walker. The two protein motors walk along the microtubules dragging organelles behind them. Kinesin walkers walk toward the positive end of the microtubule, while the dynein walkers walk toward the negative end of the microtubule. They are very strange in appearance, looking like creepy little cartoon stick figures with two arms and two legs (no head though) that fast walk along the microtubules as quickly as 100 steps per second. They use their chemical hands to grab the organelles and their little chemical feet to latch onto the surface of the microtubule.

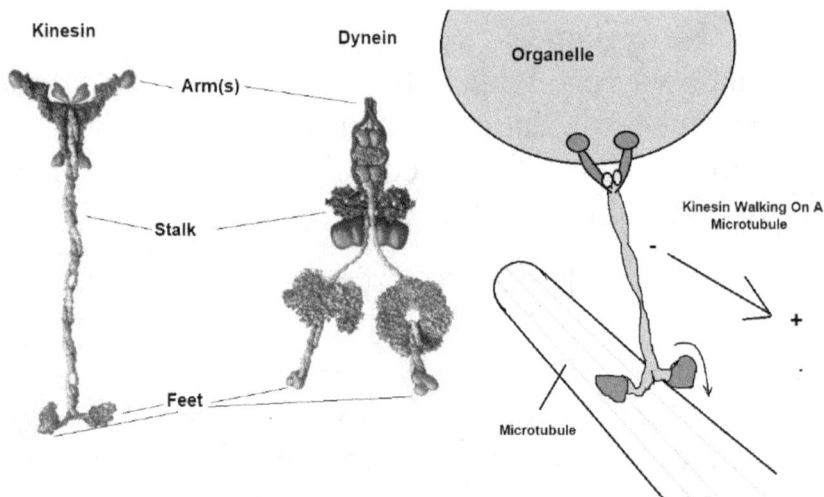

Another interesting protein motor is the flagellum motor. This protein motor bears a strong resemblance to a typical man-made electric motor, complete with stators and rotors. These motors spin the flagellum around like a boat propeller at up to 150 rotations per second, propelling the cell in the opposite direction from the spinning flagellum.

FLAGELLAR MOTOR

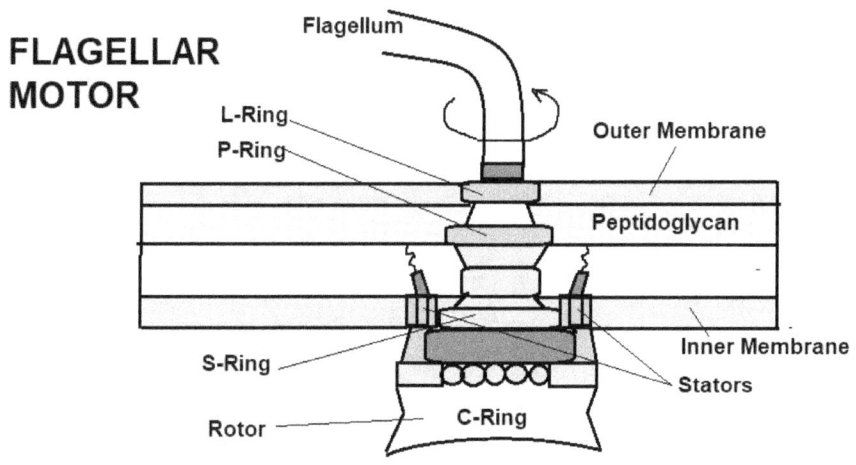

Cilia (similar to the flagellum) are powered by microtubules and dynein motors that allow the cilium to bend back and forth.

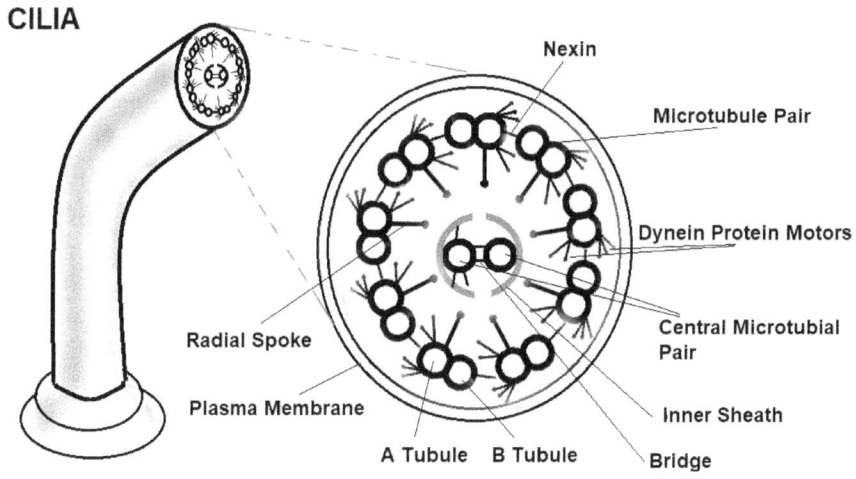

The Pilus machine is a cellular nanomachine found in bacteria. It is used by the bacteria to generate pili, which are sticky strands used for colony formation. The pilus machine is a rotary motor similar to the flagellar motor.

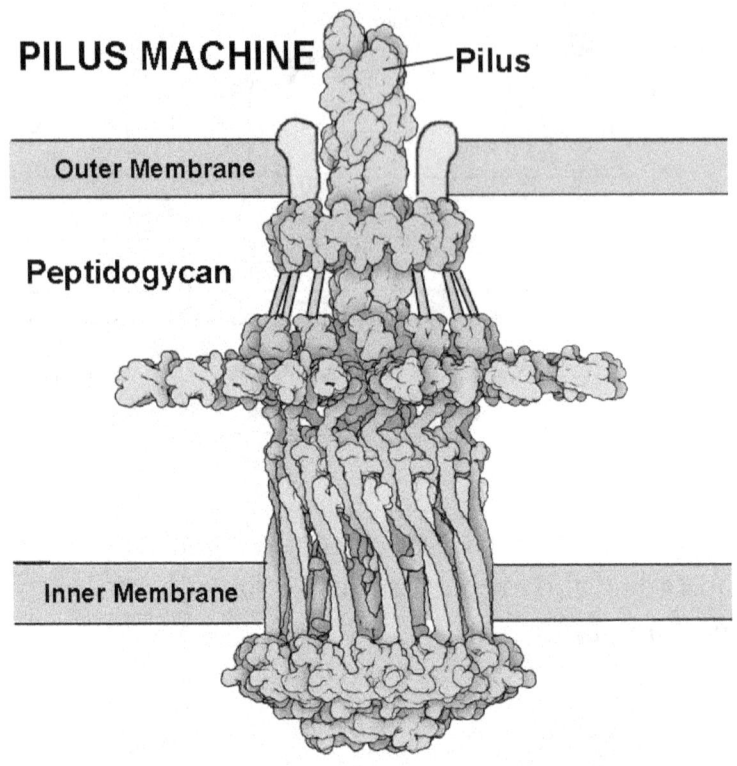

PILUS MACHINE — Pilus

Outer Membrane

Peptidogycan

Inner Membrane

There are many other types of molecular machines in cells. They are nearly all constructed out of parts made of proteins. Proteins are assembled by ribosomes using mRNA strands and tRNA attached to amino acids. The strand of protein self-folds into a part with an appropriate shape. These protein parts are then assembled and become the nanotechnology of the cell. Although many things are known about these elegant processes, we are nowhere near a full understanding of all of this complexity. One thing is certain – this new knowledge of cells makes the hypothesis that all of this complexity came about by random occurrences seem at best amazing if not magical, and at worst a proposition that begs a multitude of questions.

This chapter is an attempt to show that life itself is an amazing thing, and that the discoveries so far are just the tip of the biological iceberg. There are far more revelations to be made that quite probably will turn our current understanding of life on its head. Hopefully these findings will lead to a better understanding of the world that we live in and help lead us to a more elegant future.

The Elegant Future

Ed Nixon's Periodic Table

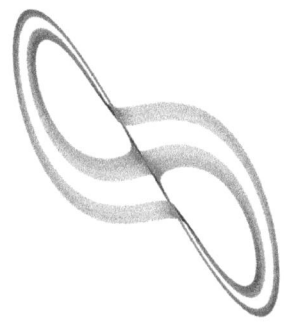

I used to help my good friend Edward Nixon (a former next-door neighbor of my wife) with his computer problems. So, I would regularly be over at his house working on his PC. I always noticed while I was there that there was a strange diagram posted on the wall over his computer desk. It was a series of spiral loops and balls that seemed to have something to do with chemistry. One day I asked him about it, and he explained it was his version of the periodic table. Ed was a geologist, so he had an interest in chemistry. When I mentioned my admiration for the diagram, he promptly rummaged through a pile of papers near his desk, pulled out another copy of the table, and handed it to me. "Here's a present for you," he said with a smile. So now that same spirally diagram adorns the wall above my own desk.

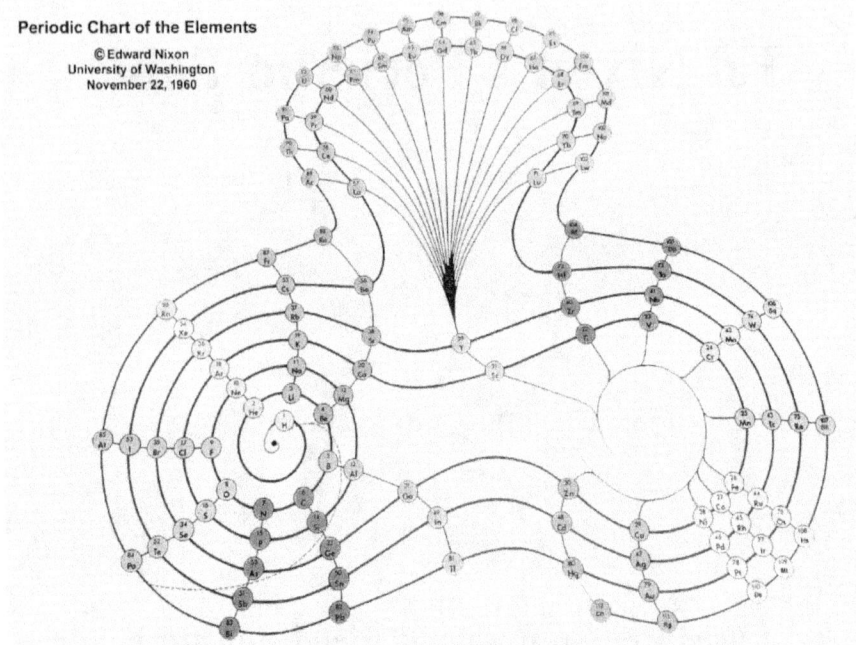

Periodic Chart of the Elements
© Edward Nixon
University of Washington
November 22, 1960

There is a simple elegance to the table. It seems to hint of a greater underlying structure to the universe itself. Later, when Elton Elliott saw the table, he proclaimed that it looked like a strange attractor. The revelation startled me. Elton was correct. It was a strange attractor. At least it certainly looked like one.

For those who have never heard of strange attractors, the definition may not be that helpful, but here goes anyway. A strange attractor is defined as the mathematical solution to a nonlinear equation for a dynamic or chaotic system. The looping pattern generated by the strange attractor solution is fractal in its form. Strange attractors exhibit stable, non-periodic behavior that is usually represented as a non-repeating (fractal) pattern in the phase space of the system.

Lorrenz Strange Attractor

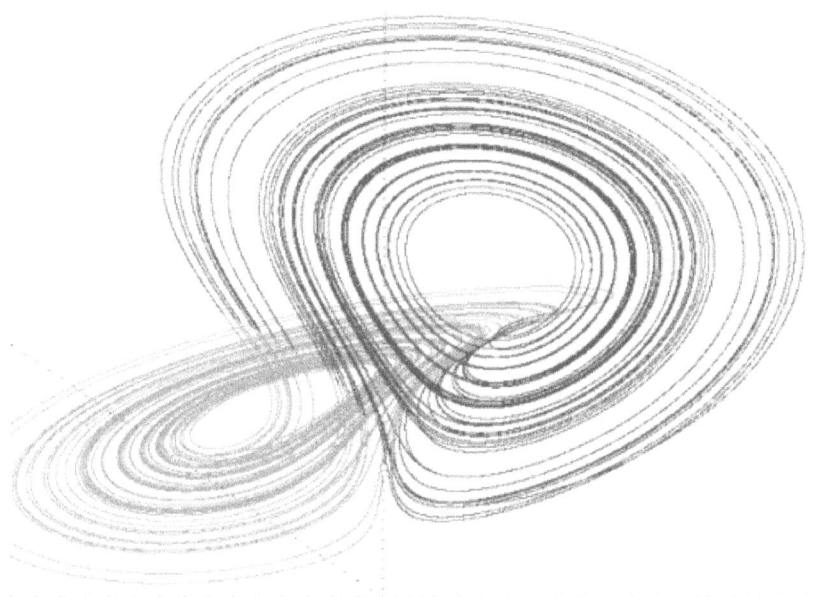

Was Ed Nixon's periodic table hinting at the underlying chaos of the universe? Was this elemental strange attractor the driving force behind the repeating fractal patterns we find in everyday life, such as snowflakes and tree branches? One thing about the behavior of a strange attractor that is clear, as you travel around the path of the solution you never quite end up in the same place that you started. Your position is always slightly shifted. This is the essence of the chaos behind a strange attractor, and it seems to be true in life as well. You can never get back exactly to the point that you started because things change slightly as you travel that looping path.

It would have been nice to get Ed's perspective on all of this, but that can no longer happen. (This was another chapter that was supposed to be written by Ed Nixon, but his abrupt departure from this world has prevented that from happening, and now the task has fallen to me.) I'm not sure what Ed would think about all of this; he might just shrug it all off, as he often did. At least he gave me a copy of this interesting discovery that he made so long ago, so it could appear in the pages of this book.

END NOTE:

ETE: I did mention it to Ed, and he muttered: "that's interesting" as he literally shrugged his shoulders. He was very modest man when it came to his accomplishments.

PART 2:

THE FUTURE IS NOW

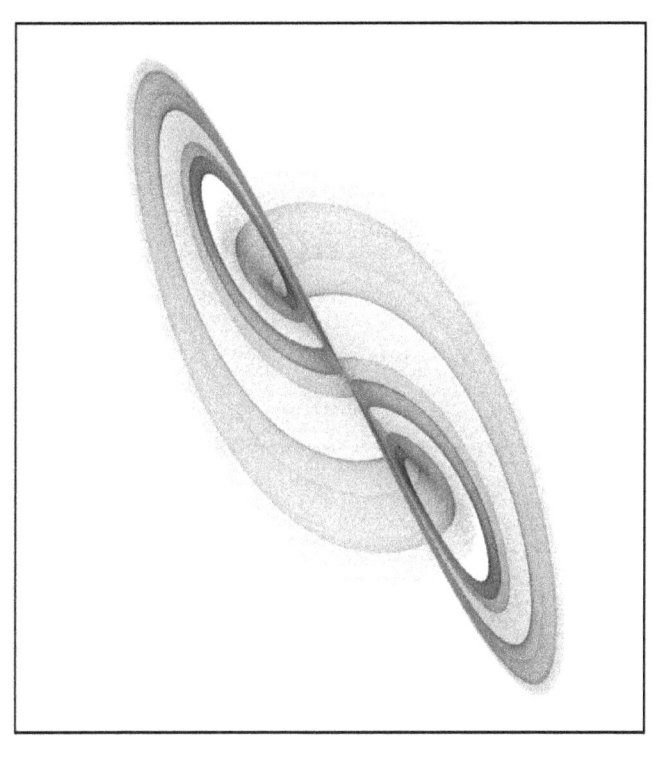

There is nothing perminant except change.

-- Heraclitus

The Hazards of Prophecy

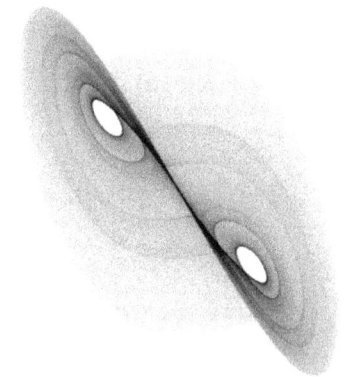

Several years ago, I (EE) was at a panel at the Locus Awards (I think it was 2013) with Gary K. Wolfe, esteemed book reviewer and literary critic. I opined from the audience that two items were going to be crucial in the future: Dyson Spheres and Donald Trump. My friend and anthology collaborator, Bruce Taylor, seated next to me, gasped as did several other folks. Mr. Wolfe stated that he did not know how to respond to such a comment.

Well, years later Mr. Trump is now President of the United States and a star about 1470 light years away (KIC 8462852, nicknamed Tabby's Star) showed changes in Lumosity that had some observers speculating that it might be a partially completed Dyson Sphere. (Professor Freeman Dyson speculated that an intelligent civilization might build a series of non-connected concentric shells around a star to collect most of its energy). So, Dyson Spheres and Donald Trump. Not bad, huh.

Of course, many in the scientific community now doubt that Tabby's Star is a Dyson Sphere. And many more in the general community, especially those attending the Locus Awards, may doubt in the veracity of Donald Trump.

Such are the hazards of prophecy.

END NOTE:

Actually, I was referencing in a cheeky way a series that I had intended to write with Doug Odell. Ironically Trump's successful candidacy put a stopper on that series. I know, the hazards, etc.

On Molten Salt Reactors and Thorium

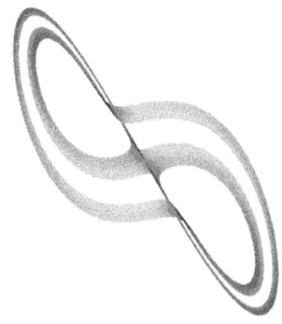

Several years ago, I was visiting with my good friend Ed Nixon at his home in Lynnwood Washington when the subject of molten salt reactors came up. He asked me what I knew about them, and I went on a diatribe about how the United States had constructed a working thorium reactor and then decided to abandon it and focus solely on Uranium powered reactors. I told him someone had made the wrong decision back then and we were now suffering the consequences of that decision. Ed looked soberly at me and said: "that was my brother who made that decision." After a few moments of uncomfortable silence, we continued talking about molten salt reactors and their potential. Ed was very interested in the fact that molten salt uranium reactors and thorium LFTR (Liquid Fluoride Thorium Reactors) were nearly identical in construction and were much safer than conventional water-cooled uranium reactors. Soon after he began a personal crusade to spread awareness of this form of nuclear power, and I would receive many subsequent emails from Ed about this subject. Ed was originally supposed to write this chapter of the book and I am quite certain that his contribution would have far exceeded my own meager words had he lived to write it.

The road to the decision Ed was referring to started in 1971 when then President Richard Nixon started a program to rebuild American nuclear power plants. At that time, Dr. Alvin Weinberg, who was the director of Oak Ridge National Laboratory, had been overseeing the world's first functioning thorium reactor. The thorium reactor was a molten salt type reactor and had been online since 1959. Weinberg strongly believed that thorium reactors should be developed as an alternative to Uranium reactors. The Nixon administration fired Dr. Weinberg in 1973 and went forward with a nuclear initiative that consisted solely of water-cooled Uranium reactors.

So why did Weinberg support thorium reactors so strongly that he was willing to lose his job over them? There are multiple reasons. One of the reasons is safety. Molten salt reactors are intrinsically safer than water cooled Uranium reactors in the fact that they cannot ever melt down. If a molten salt thorium reactor gets too hot, the freeze plug melts causing all of the fuel to drain out of the reactor, shutting it down. Thorium is also far more efficient than uranium, with one ton of thorium delivering the same energy as 250 tons of uranium. Additionally, thorium is far safer than uranium, because its only weakly radioactive (you can hold a piece of thorium safely in your hand). Thorium is four times more abundant than uranium, and easier to extract. There are currently large piles of thorium ore just sitting around waiting to be used. A liquid fluoride thorium reactor (LFTR) runs at atmospheric pressure as opposed to a water-cooled uranium reactor, which runs at 150 times atmospheric pressure. A thorium nuclear reactor produces far less waste than a uranium nuclear reactor. Another advantage to liquid salt reactors is the fact that they can be used to burn up much of the nuclear waste that is produced in a conventional reactor.

With all these advantages, why were thorium reactors abandoned in favor of dangerous inefficient uranium reactors? The short answer is that thorium reactors cannot be used to produce the materials required to create atomic weapons. The 1970s were still in the era of The Cold War, and the balance of terror had to be maintained. Pursuing thorium reactors would take some of the funding away from the kind of reactors required to maintain the nuclear arsenal.

So, what exactly is thorium? Thorium is a silvery white metal that tarnishes to black thorium oxide. Thorium is four times more abundant than uranium and is found many places, with the United States second behind Australia as having the largest thorium reserves. Thorium is also very safe. Nuclear fission will not occur in thorium alone. Uranium must be used as a primer to start the nuclear reaction.

The Elegant Future

The molten salt reactor is the best means to use thorium as a nuclear fuel. The basic design of such a reactor is shown below.

MOLTEN SALT REACTOR

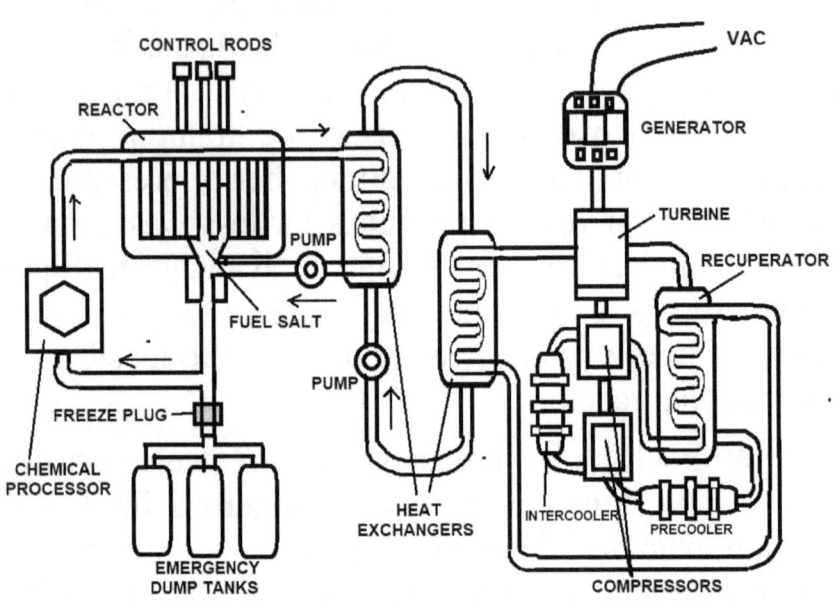

The main advantage of this reactor design is its safety. If something goes wrong and the reactor gets too hot, the freeze plug beneath the reactor melts away and drains all the fuel out of the reactor core and into holding tanks beneath the reactor. This completely deactivates the reactor. In a conventional water-cooled reactor, if the cooling system fails the reactor melts down causing a radioactive catastrophe.

Because of all the positive qualities of thorium LFTR reactors, many countries are trying to develop the technology. In 2013 China partnered with Oak Ridge to pursue the development of thorium reactors. India has large deposits of thorium and is eager to exploit this resource and has set the goal of getting 25% of its energy from thorium powered nuclear reactors. Israel has been working with Brookhaven National laboratories to try to develop thorium LFTR reactors. Norway's Thor Energy company is working with Nobel laureate Carlo Rubbia to create a proton accelerator-based thorium reactor. The United States, the creator of the first working thorium reactor, is quietly working with Russia and China to develop thorium reactor technology.

It's easy to see why Ed Nixon, along with many others, became a proponent of thorium and molten salt reactors. If nuclear fission reactors are going to continue to be part of our future, thorium LFTR reactors seem to be one of the most elegant solutions for that future.

The Elegant Future

Miniature Suns: The Promise of Fusion Power

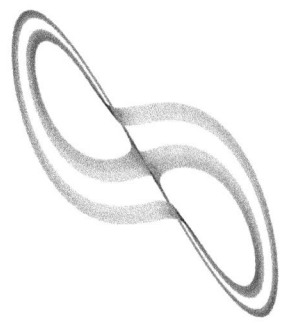

Ninety-three million miles from the Earth floats a giant natural fusion reactor known to us as the Sun. The titanic crushing forces at the core of this massive plasma ball are so great that atoms of hydrogen are fused together to form helium. The pressure at the Sun's core is more than 250 billion atmospheres (250 billion times the atmospheric pressure on Earth at sea level), and the temperatures are in excess of 15.6 million degrees Kelvin. The daily energy output of the Sun is on the order of 384.6 yotta watts per day (that's yotta not Yoda), which is 384.6 million million million million watts. Mankind is currently using approximately 15 terra watts per day, which means the daily output of the Sun is 15 quadrillion times the amount of energy that mankind requires. Of course, only a tiny fraction of this vast blast of energy actually strikes the Earth, it being such a small target. Even so, enough energy strikes the Earth in one hour to power the world economy for an entire year.

The most common fusion reactions in the Sun are as follows:

$$_1^2H + {_1^2H} \rightarrow {_2^3He} + {_0^1n} + 3.27 MeV$$

$$_1^2H + {_1^2H} \rightarrow {_1^3H} + {_1^1H} + 4.03 MeV$$

$$_1^2H + {_1^3H} \rightarrow {_2^4He} + {_0^1n} + 17.59 MeV$$

$$_1^2H + {_2^3He} \rightarrow {_2^4He} + {_1^1H} + 18.3 MeV$$

In all cases two lighter atoms are forced together to form a single heavier atom, and excess energy is released.

For many decades countless scientists have tried to replicate the Sun's fusion processes in laboratories without success. These same scientists have been telling us that cheap fusion energy is just around the corner for the last several decades. Many different methods have been tried, but the task has proven to be much more daunting then was originally anticipated. There have been four major methods used to attempt to imitate the atomic transformations at the center of the Sun. They are the Tokamak Confinement reactor, the Inertial Confinement reactor, the Electric Pinch Confinement reactor, and the Inertial Electrostatic Confinement reactor.

The most popular method of fusion to date has been the Tokamak reactor, which was originally developed by Soviet physicists Igor Tamm and Andrei Sakharov in the early 1950s. Currently several nations are collaborating on a giant Tokamak reactor in France. The project is called ITER, and the nations involved are the United States, the European Union, India, Japan, China, Russia, and South Korea. So far nearly $15 billion has been spent on the project, but the reactor is still under construction. The completed project is expected to cost in excess of $25 billion. Will the ITER project result in a functioning fusion reactor? Don't hold your breath. Mankind has been playing with Tokamak reactors for nearly 70 years without success (success being more sustained energy produced than is required to run the reactor, or what's known in the fusion business as a Q factor greater than one). Many prominent physicists believe the Tokamak is a bad design. Plasma Physicist Dr. Nicholas Krall once joked, "We spent $15 billion dollars studying tokamaks and what we learned about them is that they're no damn good."

The Elegant Future

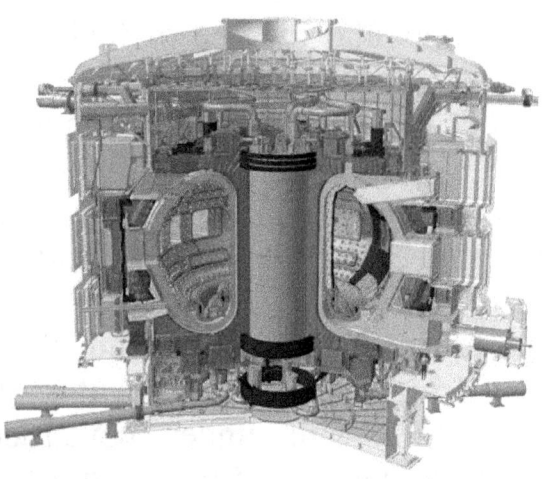

Diagram of the ITER Tokamak reactor.

Another fusion method that is being explored is Inertial Confinement. In this method, multiple high-powered lasers are shot at a pellet of fuel. The method is called inertial confinement because it uses the fuel pellet's self-inertia in the confinement process (the fuel pellet's own inertia prevents the deuterium atoms from flying apart, confining them long enough for fusion to take place).

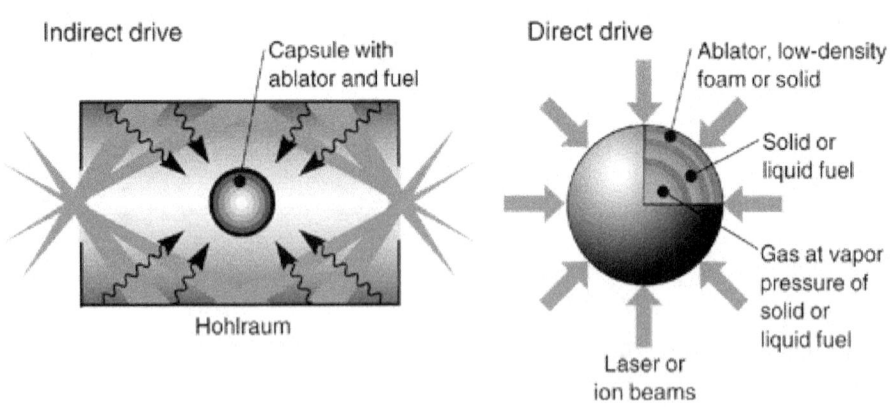

So far, Inertial Confinement has not provided mankind with a Q >1, but the NIF (National Ignition Facility) did manage to provide *Star Trek Into Darkness* a complicated backdrop for one of the film's sets.

Electric Pinch Confinement (Z-Pinch) uses a natural physics phenomenon called electromagnetic pinch. Electromagnetic pinches can occur naturally when high magnitude electrical discharges occur such as in lightning bolts, the aurora, and even solar flares. An electromagnetic pinch can occur in any conducting material such as cans, metal pipes, or plasma in the case of Pinch Fusion reactors. The University of Washington's ALPHA project (Accelerating Low-Cost Plasma Heating and Assembly) is a Zeta Pinch reactor.

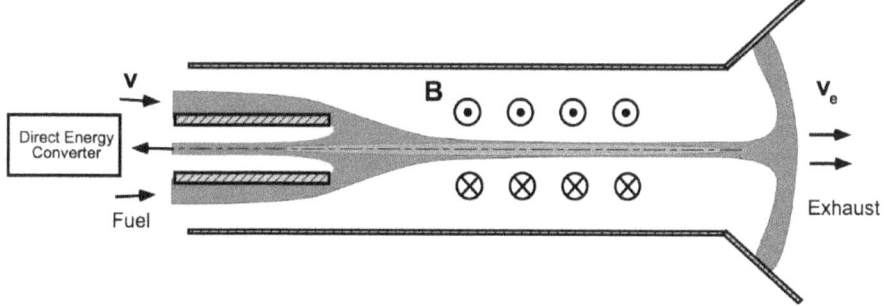

UW's Z-Pinch Reactor

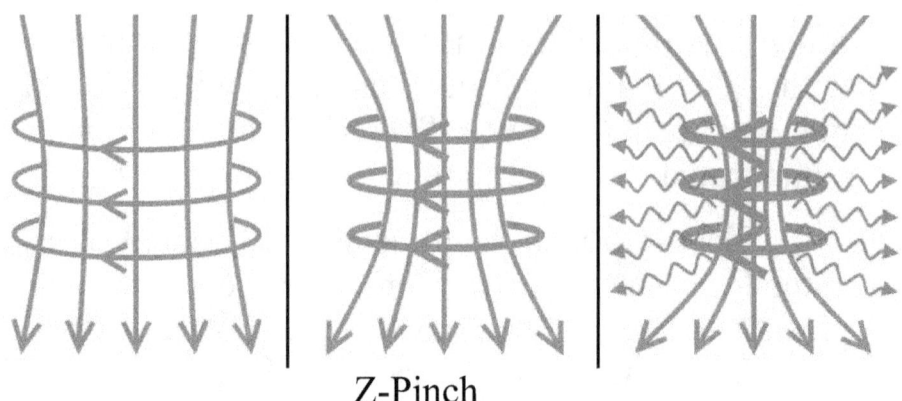

Z-Pinch

The plasma current self generates a magnetic field which in turn compresses the plasma causing energy to be released.

The final and perhaps most promising fusion reactor is the Inertial Electrostatic Confinement reactor. The Bussard Polywell reactor is such a device.

Bussard Polywell

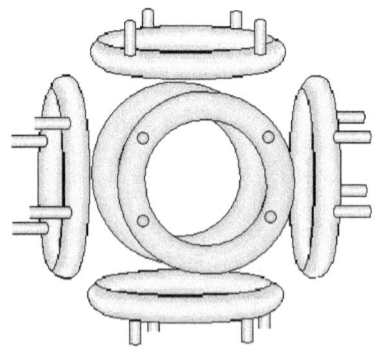

In the Bussard reactor, six electromagnetic coils are arranged into a polyhedron. This results in a magnetic null point at the center of the arrangement. Electron guns are used to shoot electrons into the device's magnetic fields, and they become confined. This results in a virtual cathode which causes hydrogen atoms to fall into the potential well and fusion occurs at the null point. Korean scientists believe they will achieve fusion which such a reactor 30 years ahead of the ITER project primarily due to the relative simplicity of the reactor and the fact it will be tens of billions of dollars cheaper to build.

Whichever reactor design succeeds is irrelevant. The point is many highly intelligent people are working on achieving nuclear fusion. (I am personally cheering for one of the smaller groups on shoestring budgets to succeed.) Fusion energy will eventually become a reality and that could mean unlimited sources of energy.

The Elegant Future

The Cell Phone in Your Pocket

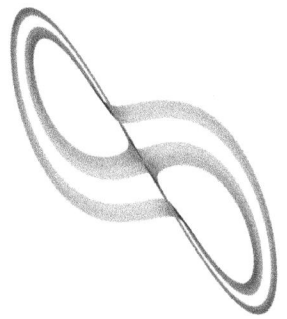

Most people carry a miracle of technology in their pocket, without ever considering its incredible capability or its potential dangers. The device is called a cellular telephone, a lackluster name for a device capable of so many things that it boggles the mind to consider them all. The lowly cell phone is a wireless telephone capable of contacting every other human being on the planet (that owns a similar device), and nearly *everyone* owns such a device now. I recall several years ago, being in Vietnam in the middle of nowhere standing near a rice paddy. I watched a woman wearing a *nón lá,* (or conical sun hat) ride up on a water buffalo from the middle of the nearest paddy. She hopped from her mount, pulled out a cell phone, and began conversing with someone. That was nearly five years ago.

The Elegant Future

Today's cellphones are a multifunction miracle. My cell phone is also a high resolution still camera capable of capturing breathtaking photos in almost no light. It also contains a 4K video camera that can be used to capture videos of motion picture quality. It has a voice recorder and can play literally thousands of music files. It is a clock and a calendar. If I forget my wallet, I can make purchases with just the phone if I have the proper application. I have access to the internet through the device which essentially connects me with all the knowledge of mankind. I can send text messages and images/videos taken with the phone to anyone almost instantly. If I get lost, it has a GPS and can tell me exactly where I am to a few hundred feet. With a Bluetooth wireless connection (named for Harald Bluetooth, a former King of Denmark and Norway) and a few attachments, it can be used to make medical measurements of such things as pulse, blood pressure, and other vital signs. It is a pocket computer and can run myriads of apps that formerly required a bulky desktop box. And of course, it can also serve as a flashlight. A modern cell phone makes the tricorders from the original Star Trek series pale in comparison.

There is a dark side to these tiny miracles of modern science. Cellphones can connect us to everyone in the world, but they are actually responsible for causing disconnection and loneliness in many people. Riding in elevators, I am often the only person who is not staring at a smartphone. In a darkened theater waiting for a show, it is dead quiet except for the multiple glows of tiny screens mesmerizing the audience. People ignore each other in favor of their phones. According to an AARP study, there is a loneliness epidemic. In the study, 43 million adults, 45 and older, across the U.S. were found to suffer from chronic loneliness, a disorder that can cause depression, and potentially death. Cell phones play a significant role to augment this situation.

These effects are particularly harmful to children. According to American psychologist Jean Twenge, children are spending too much time on their cell phones, which is placing their mental health at risk. Surveys show that today's kids have greater anxiety than ever before, and depression rates are rising rapidly, along with suicides.

Another problem with cell phones is even more diabolical. Your cell phone doesn't just transmit when a call is made, it is constantly pinging out its location. There are apps available that can allow another person to track the location of your cell phone, and even track your phone's call log and social media activities. Some of these applications are even free to download. If you have a cell phone, other people could be watching your every move. Although it is possible to go to the phone's setup and disable location services, the only sure way to prevent someone from tracking your phone is to turn it off.

In addition to privacy concerns with cell phones, there are also health and safety concerns. A cell phone is a radio transmitter. Its power output automatically varies depending on the repeater tower availability, but it can be as high as 1 watt (a microwave oven puts out about 1000 watts). Short exposures to this tiny amount of energy is harmless, but long-term exposure is another matter.

In November 2018, the National Toxicology Program released a report concerning a study of the effects of radio frequency radiation on rats. The NTP discovered such exposures led to decreased body weights, and lower birth rates for rats exposed to the same type of radiation emitted by cell phones. They also discovered increased numbers of cancerous and precancerous tumors in the hearts, brains, and various glands of the animals. The NTP did a similar study with mice and found similar results. In the mice, tumors tended to be located in the livers, skin, and lungs. In both studies, female animals fared slightly better than male animals. So, cell phones radio emissions were found to be dangerous to rodents causing "biological effects relevant to carcinogenesis." In addition, in 2014 the International Agency for Research on Cancer concluded that RF radiation from cellphones was "possibly carcinogenic to humans." Both the NTP and the IARC are quick to say that the effect on human's could be very different from the effects on test animals, which means the effects of RF on humans could be less dangerous, or they could be more dangerous. Both organizations want to do further research.

All this controversy hasn't stopped cell phone companies from developing even more powerful phones. 5G cell phones are already upon us. These new phones emit even more radiation at higher frequencies to reduce latency and vastly increase bandwidth. According to Qualcomm (a manufacturer of cell phone integrated circuits): "5G will be able to boost capacity by four times over current systems by leveraging wider bandwidths and advanced antenna technologies."

So, cell phones are not risk free. To reduce the risk the following guidelines are suggested:

1. Turn off your phone if you aren't using it and are not expecting calls. (You can always check your phone log and call people back.)
2. When you're inside, don't keep your phone on your person, but place it somewhere where you can hear it ring.
3. Use a headset (or put the phone on speaker) when talking on the phone for long periods and keep the phone at least two feet away from you.
4. Don't use the phone to cruise the internet if there is a free PC available.

The cell phone is a fantastic invention, but use it in moderation and with caution so you can live long enough to enjoy the elegant future that lies ahead.

The Elegant Future

Someone is Watching You: The Loss of Privacy

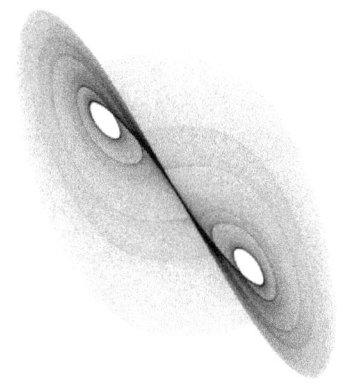

We have all felt it. That moment when you think you're alone, maybe walking in the woods, or sunning yourself in the supposed privacy of your back yard, and then you get that prickling sensation on the back of your neck. You feel you're being watched by someone or something. You whirl around, and no one is there. Or maybe they are. In the words of the Alan Parsons Project song:

> "It's almost a feeling you can touch in the air
>
> You look all around you, but nobody's there
>
> It's been a long time now since you've been aware

That somebody's watching you."

In Seattle in a high-rise condominium tower, a woman was cleaning her house in her lingerie and she noticed a drone hovering outside. When she tried to discover who owned the drone it turned out it was owned by a real estate company. No word about whatever happened to the video taken of her cleaning. Another real estate video taken by a drone showed a woman sun-bathing in her back yard. She sued the company, and reportedly settled out of court, but the video had already made it to the internet.

Observation. There are cameras everywhere. That's so true for much of the industrialized world right now. But at least you can see them. Some of them. The most insidious are invisible and it will only get worse as information technology becomes increasingly miniaturized and as the computer, like water and sewer pipes and electrical wiring before it, recedes into the walls of our houses, offices, factories, and schools. Eventually all of the cameras will be invisible, virtually undetectable. At least the cameras are only outside. You won't have to worry if you're inside your home or apartment. Right?

Not so fast. Now comes word that export controls over so-called "smart dust" are one of the major items on the agenda of United States federal regulators. Smart Dust is the latest buzzword describing extremely small surveillance machines. The formal definition of smart dust is *a collection of numerous tiny microelectromechanical systems that together form a computer. These minuscule devices are light enough to remain continuously suspended in the air and are used mainly for information gathering.* Imagine clouds of these microscopic eave's droppers, just floating in the breeze. These highly miniaturized cameras and other recording devices can enter public buildings, private businesses, schools, locker rooms, people's homes almost completely undetected. Think of that. They'll roam into every room, living rooms, kitchens, bedrooms, bathrooms. You won't even know they are there.

While as unappetizing as home invasion is, the loss of privacy is occurring in two other major areas. First, most importantly is commercial espionage, business and trade secrets. Second, but no less important, just not as frequently, is espionage in the defense arena. Sub-contractors are more vulnerable than the military itself. (Note that we have not mentioned cybersecurity; that will be handled in a separate chapter). See also the chapter on 3D Travel.

What to do about it?

Well, help support technology that values privacy and freedom.

Today it's mostly cybersecurity, data encryption, etc. Tomorrow it may well be machines that warn against and prevent video intrusion.

The battle for our privacy has just begun.

The Elegant Future

Genetic Engineering: CRISPR CAS9 -- the Magic Scissors

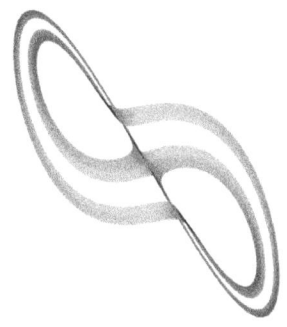

Genetic engineering was once a hit and miss proposition. Only rarely was the intended result achieved. Most of the time experimental genetic manipulation resulted in failure. Then in 1987 at Osaka University in Japan, Dr. Yoshizumi Ishino and his colleges accidentally cloned a unique repeating sequence of genes along with their intended target gene, the Inhibitor of Apoptosis. This mysterious sequence turned out to be part of a very special marker sequence which was later dubbed CRISPR. Later in 1993 Dr. Francisco Mojica, a Spanish microbiologist at the University of Alicante, identified the same sequence in his own experiments. He coined the sequence as CRISPR for 'Clustered Regularly Interspaced Short Palindromic Repeats.' Once again in 2002, Ruud Jansen of Utrecht University in the

Netherlands found the same CRISPER sequences in different organisms in distinct spacings which varied depending on which organism the sequences were found in. It was later established that these CRISPR sequences in concert with an enzyme known as Cas9 (CRISPR-associated enzyme number 9) created a sort of 'Magic Scissors' which would enable geneticists to precisely cut and splice genetic material. According to David Baltimore, a Nobel Prize winning biologist, the discovery of CRISPR Cas9 techniques are a "monumental moment in the history of biomedical research."

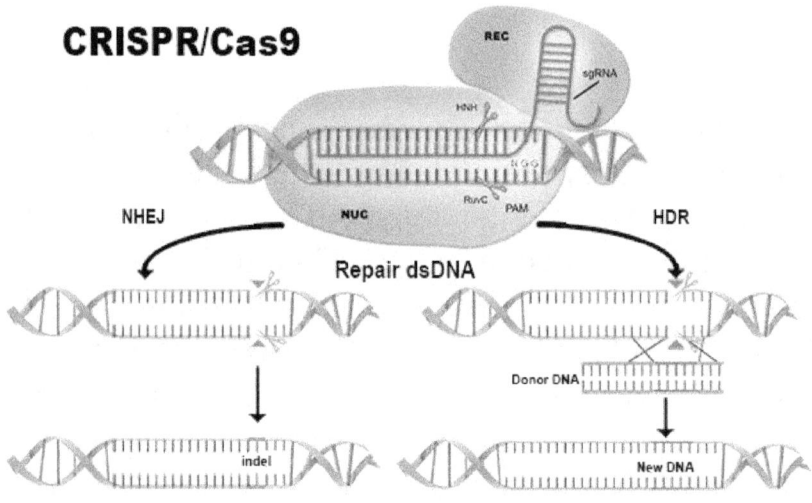

CRISPR Cas9 technology has led to a series of breakthroughs in genetic engineering in a wide range of fields. One such field is farming and agriculture. Genetically Modified organisms (or GMO) are now commonplace. Nearly 85% of the food we eat contains GMO products. Nearly all the corn grown in the United States has been modified. Soy is one of the most genetically modified foods, with certain varieties of soy modified to produce omega-9 fatty acid in hopes of reducing bad cholesterol. Squash, alfalfa, canola, and sugar have also been genetically modified either with CRISPR Cas9 or other gene-splicing methods. Cows have come under the magic scissors as well, in order to increase milk production, and to make them grow faster.

Medicine is another area where the new gene-splicing capabilities are creating breakthroughs. CRISPR techniques are being used to fight cancer by using genetically engineered CAR-T cells (Chimeric Antigen Receptor Thymus Lymphocytes – a type of man-made white blood cell) to fight tumors. CRISPR has been used to fight AIDs by extracting HIV-1 DNA from infected human cells and further modifying the cells so they cannot become re-infected. New techniques have created self-destructing disease causing bacteria. CRISPR is being used to reduce the possibility of miscarriages, and correct genetic mutations in human embryos. It may even be possible to eliminate malaria by creating malaria resistant mosquitos. (No malaria but more healthy mosquitos?)

CRISPR Cas9 has also been used for silly things. Exotic items such as glow in the dark pets and houseplants have been created. Harvard scientists are currently using CRISPR to recreate a wooly mammoth by splicing mammoth DNA into elephant embryos. Another group is trying to turn chickens into dinosaurs. Chinese scientists at the Guangzhou Institutes of Biomedicine and Health have used

CRISPR to create a muscle-bound beagle that would put Arnold Schwarzenegger to shame.

All these amazing advances are not without risk. Due to the potentially holographic nature of information storage in DNA, editing genes using CRISPR can result in unintended consequences. A simpler way of putting this is that genes can serve multiple purposes. Changing a gene to improve a characteristic may cause other unforeseen problems. These dangers are known as off-target effects. Off target effects are essentially the unforeseen results of gene editing. For example, editing a gene to treat one medical issue may activate a cancer-causing gene. Or editing may cause something known as translocation which can cause two different chromosomes to connect. Myeloid leukemia is a fatal blood disease caused by translocation. ATR-16 syndrome is another rare genetic disorder caused by translocation. Affected individuals have symptoms which include stunted intellectual ability, clubfoot, reduced head circumference, and alpha thalassemia, a blood disorder that causes reduced levels of hemoglobin, the oxygen carrying molecule found in red blood cells.

Despite the dangers of gene editing, the CRISPR Cas9 splicing techniques holds great promise for the future. ExxonMobil has used CRISPR to create a new biofuel generating algae. Gene splicing could modify humans to be better athletes or become better scientists by increasing intelligence. (Would edited athletes be allowed to compete against genetically standard humans?) Dr. George Church of Harvard Medical

School is working to use gene editing to reverse aging in dogs and eventually humans. Such gene editing could be used to reset the aging clock in human cells and allow humans to have indefinitely long lifespans. Gene editing could eventually create entirely new forms of life, for better or worse.

The genetic genie is already out of the bottle and is not going anywhere. The test will not be whether humanity uses these newly found powers. The real test will be what wishes this genie is asked to grant and the subsequent consequences to humanity.

The Elegant Future

3D Printing and Instantaneous Fabrication

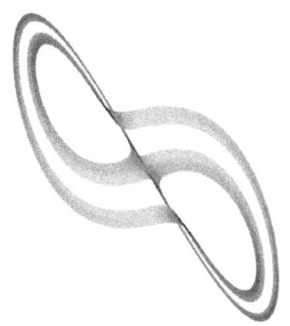

In the mid-late 1990's Dr. Arlan Andrews, who was then working at Sandia National Laboratories, showed the authors of this book a radically new device that could print 3D objects. The machinery filled a room, sounded like a jet engine, and cost millions of dollars. The monstrous device could create tiny plastic vases too small to be of practical use (although they probably could have modified it to make golf tees). I remember watching as the device worked its scientific miracles. It was as if the little vases were materializing before our eyes like a Star Trek transporter. Now, more than 25 years later I have a 3D printer sitting on my desk and it cost less than $1000. It can print out an amazing assortment of objects. I've used it to fabricate specialized parts for experiments, and to create plastic models of objects from my books. Although this newer printer is still a real pain in the butt to set up (mostly print leveling problems and clogged nozzles), once you finish pulling your hair out (very little left now), and cursing at it, and finally getting it just right, it can do an amazing job of fabrication. Suddenly every 3D object I've ever created can be called into reality with a little patience, and a big enough spool of plastic filament (it can take days to finish a large complex print). Another useful aspect is the ability to create different sized models of the same object. A big intricate print can be checked out by initially printing a smaller version for examination before scaling it up to full size.

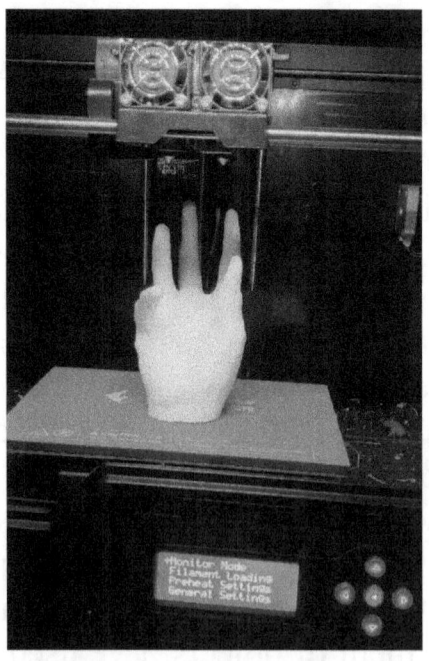

3D printed hand.

The same human skull printed in 2 sizes.

Three-dimensional printing has a convoluted history. The first generally accepted 3D printing method was conceived by a man named Hideo Kodama in 1980. Dr. Kodama designed a rapid prototyping technique which involved using a laser beam to cure plastic resin into a solid. The device was never completed due to a lack of funding. Several others attempted to create three-dimensional printing techniques and also failed. A functional printing technique was finally realized and patented by Charles Hull in 1984. Hull dubbed the process stereolithography, and started a company named 3D Systems in 1986. That company is still going strong today and has created multiple technological advances in the field of 3D manufacturing, including stereolithography printers, selective laser sintering, color jet printing, surgical simulators, direct metal printing, and digital light printing (DLP).

The Elegant Future

Today there are an amazing array of rapid three-dimensional printing technologies available, and the processes are getting faster and more refined with each passing month. In addition, the cost of these devices is dropping as well. Devices that used to cost hundreds of thousands of dollars can cost less than high-end cameras. These printers can also print with a variety of materials. The printer on my desk is capable of printing with PLA plastic, ABS plastic, T-glass, and composite fiber material such as copperFill, brassFill, and more. Newer industrial class printers can print with glass, wood, and metals such as copper, brass, aluminum, and steel. Steel parts must be sintered to remove binding agents and make the parts pure metal, but the printed objects are equivalent in strength to machined parts and cost only a tiny fraction of what a machined part would cost. For example, a 3D printed rotor costs $14.07 whereas the same machined part cost $1607.58. Printers can also combine different materials in the printing process, such as metal and plastic. In addition, 3D printing allows parts to be fabricated with geometries that cannot be created by conventional machining or molding techniques.

Now even human tissue and organs can be 3D printed. Alginate or Fibrin polymers, enforced with nanocellulose, and integrated with the appropriate cellular adhesion molecules can be used to print an organ scaffold which can then be seeded with human cells that can grow and multiply throughout the scaffold to become the desired organ. Since the human cells can be harvested from the recipient for the constructed organ, organ rejection can be largely eliminated. With a sufficiently detailed printing technique, pre-existing vascularization can be directly printed in the organ. This ensures proper oxygenation, nutrient supply, and waste removal for the printed tissue. Simple organs such as skin, bladders, and blood vessels are already being printed, with experts believing that complex organs such as hearts and kidneys are less than a decade away from reality.

So how will 3D printing and other rapid manufacturing techniques affect the future? When 3D printer technology is sophisticated enough, many items need not be shipped. Instead only the 3D patterns of the object need to be transmitted over the internet and the resultant object created at the destination point. Of course, if you are handy with one of many excellent 3D modeling tools available, you can bring your design to the 3D fabrication station and have them print it out to your specifications. This is already largely true today, but printing costs will continue to fall making this technique more accessible than ever. With 3D printing, if you can imagine it, you can hold it in your hands.

Rapid three-dimensional fabrication is not limited to small items. There are now large-scale 3D printers capable of printing houses. The printer is essentially one or more giant robotic arms that can extend, retract, and elevate while extruding a concrete-like paste. Multiple extruders allow insulative foam to be printed between the outer and inner walls while the house is formed. Using this technique an entire structure can be created in 24 hours or less. The printing also greatly reduces the cost of constructing houses. ICON, an American construction company recently created a 650 square foot 3D printed house in Austin Texas for a cost of about $10,000. The company believes that the cost of such a house can be eventually dropped to about $4000.

With the steady progress that has occurred with 3D printing and rapid manufacturing techniques, one can only imagine what the future could hold. If organs for transplants can be successfully printed and costs dramatically lowered what is the next step? What about an extra back up heart for those that want system redundancy to prevent a single point for catastrophic failure? Just have a second heart printed out and installed so if your original fails the other keeps the blood circulating. What was once certain death is now a minor inconvenience. What about an extra kidney just to be on the safe side? If we can print out organs will we be eventually able to print out entire animals? (I am reminded of the scene from the movie Fifth Element where the main protagonist is recreated by a three-dimensional fabrication process after being killed in a crash. A silly movie, but I found that scene intriguing.) Just get a detailed scan of your cat as a backup. If it gets run over by the Amazon delivery van, just print out a new one. If the process is detailed enough, will the new cat get the old cat's memories? Will the 3D printed backup of your dog run to you wagging its tail? Forget conventional animals, what about printing out new customized animals by combining patterns from multiple sources. This starts getting really weird fast. If it works for animals what about humans? Suddenly ethical questions emerge. Be sure to back up yourself and your children in case the unthinkable happens.

The Elegant Future

The Potential and the Problems of a Truly 3D World

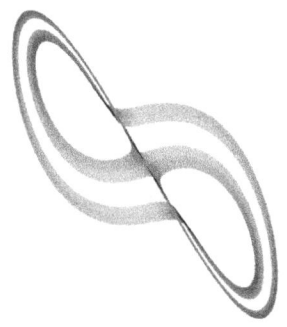

Since the beginnings of human history, mankind has lived in a world that is essentially two dimensional. Traveling through the world is restricted to a plane of directions with traveling upward prevented by gravity and traveling downward stopped by the ground beneath our feet. The use of animals, either to ride or pull wagons, allowed us to get places faster, but still in restricted dimensions. Bicycles, trains, and automobiles got us places even faster, but were still limited to the same plane of motion. When air travel was possible, mankind soared into the skies and the two dimensions were finally breached, or were they? Aircraft travel is restricted to flying along predetermined courses from one airport to another. Then the traveler must return to two dimensions using automobiles and trains again. Even the air travel portion could be classified as two dimensional once the plane has ascended to its cruising altitude. Aircraft are still restricted to a narrow region between the ground and the edge of space. So, in a large sense, the advent of airplanes was simply another means of making traditional two-dimensional travel more efficient and faster.

This 2D legacy is beginning to change. The helicopter is an early example of this. The helicopter is an offshoot of an earlier device called an Autogyro (or gyrocopter). The gyrocopter has an unpowered rotor that spins by autorotation. Unfortunately, this means it still requires a runway to take off. The Focke-Wulf FW 61 was the first functional powered helicopter which was created in 1936 by Henrich Focke. Igor Sikorsky created the first mass production modern helicopter in 1942 with his XR-4. The helicopter's Vertical Take-Off and Landing capabilities (VTOL) has allowed them to be used in ways normal airplanes can't. A helicopter can be landed next to (or on top of) a hospital, allowing for quick transport of critical patients to, from, or between medical facilities. Helicopters allow police to hover over neighborhoods, allowing them to track down dangerous criminals. Helicopters have allowed humans to wade into the shallows of the 3D world, but like airplanes they are usually restricted to airports and specific landing areas, and are complex machines requiring specialized training to fly them.

In 1991, two companies, Sony, and Asahi Kasei released the first commercial lithium ion batteries. These batteries were powerful, but light enough to allow the creation of small electric powered flying machines. The first fully electric unmanned drones were born. Although powered unmanned drones are not a new concept (the Hewitt-Sperry Automatic Airplane called the bug was created in 1918), gasoline powered engines were required for flight. Now batteries exist that are light enough that small flying electric unmanned aerial vehicles can be a reality (UAVs). Today flying electric drones are everywhere. By coupling VR goggles to a drone with miniature video cameras one can at least virtually experience a truly 3D reality.

This new 3D world doesn't stop with UAVs. Several companies have created manned vehicles using drone technology that will allow the average person to break the surly bonds of Earth. Surefly aerospace company has recently unveiled an electric drone capable of flying two people (or one person and cargo) up to 112 kilometers (70 miles). A Chinese company, EHang, has a similar product out. Uber has a new project called Uber Elevate which intends to use drones to ferry people about. Both Airbus and Bell Helicopters are working on their own version for manned drones. Most of these drones will have autonomous capabilities which means passengers won't need to have a pilot's license to use them. The Dubai police force is purchasing drones that are essentially flying motorcycles from a Russian company called Hoversurf. Drones are only the beginning. One can imagine a future were a person could strap on (or step into) a relatively small device that would enable them to fly like Superman.

This creates some serious security concerns. A fence or wall will not thwart anyone who has access to this technology. Any open window or unsecured balcony no matter how high could provide unfettered admittance for a creative criminal who can be ambulatory in all three dimensions. Moreover, higher windows may actually be more desirable in this scenario as they will be far from the street level and more secluded. Rooftop ventilation systems become front doors for vertically mobile malefactors. Terrorists could deliver IEDs using unmanned drones, or simply toss them down from their hovering Harleys. New security measures will be required. Security cameras will need to look up as well as down. Security guards will need to fly instead of drive. Standard home accessories will need to include a high-tech aerial warning system that would detect unwanted individuals who trespass too close from above.

There is a lighter side to this predicament. If this levitation technology continues to advance, sports will never be the same either. One can imagine manned drone races around an aerial arena. Skydiving without a parachute may be the next thing, where people bail out of airplanes and power their way to a safe ground landing. Of course, the airplane may be superfluous with powerful enough energy sources. Perhaps then the new sport will be to skydive up to the airplane instead.

What about 3D soccer? High tech hovering brooms could make J.K. Rowling's Quidditch a reality.

Eventually technology may advance to the point where an average human can own a small affordable vehicle that not only could allow them to fly anywhere on Earth, but to also travel beyond the Earth and out into space as well (Jetsons like technology). This would allow for space travel on a massive scale. Going to the Moon would be a weekend trip. This makes me recall a conversation I had with the late Timothy Leary, who believed that this extra dimension of mobility would change (or in his words expand) people's minds resulting in a fundamental change in humanity itself. Whether or not 3D mobility ultimately changes us is a question that can't be answered here, but it will certainly give us an entirely new perspective on our world and everything in it.

The Elegant Future

The Super Encrypted Future

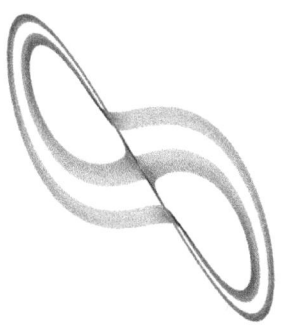

We live in a world that is run by data. Data controls all of the transactions we make at the grocery store, at the bank, at the doctor's office. Online purchases are nothing more than the exchange of data over the high-speed communication lines of the internet. If anything happens to the secure transfer of this data, the entire system is put in peril. Sensitive personal data must be protected to prevent electronic funds from being hijacked and lives jeopardized or potentially ruined. Data is protected by encrypting it. Encryption is basically a means of scrambling the data, so a third party is unable to access it. There are many types of encryption out there, but only two national standardized forms. They are the DES (Data encryption Standard) and the AES (Advanced

Encryption Standard). Unfortunately, both forms of encryption have been compromised to some degree, if not outright broken. In addition, the exponentially growing power of computer systems will inevitably require a far more powerful form of encryption.

Before today's encryption issues can be properly discussed, it is useful to trace the origins and development of encryption systems. The Spartans were the first to use a form of encryption in 600 BC. The encryption consisted of a leather strap wrapped around a wooden rod of a very specific diameter. The message is written on the leather and when the leather strapped is unwrapped the letters on the strap become meaningless. Only a person with the correct size rod can re-create the message.

The next developments in encryption came from the Arabs. Arab philosopher Al-Khalil ibn Ahmad al-Farahidi (717 AD – 786 AD) wrote a book called *The Book of Cryptographic Messages*. Arab mathematician Abu Yusuf Yaqub ibn Ishaq aṣ-Ṣabbah al-Kindi wrote another encryption book in 800AD called *Risalah fi Istikhraj al-Muamma* (*The Manuscript for the Deciphering of Cryptographic Messages*), which contained cryptanalytic techniques, polyalphabetic ciphers, cipher classifications, Arabic phonetics and syntax, and a section about frequency analysis. Later, Egyptian scholar Shihab al-Din abu l-Abbas Ahmad ben Ali ben Ahmad Abd Allah al-Qalqashandi al-Fazari (al-Qalqashandi for short) (1355 – 1418) wrote a section about encryption in his 14-volume encyclopedia set.

In the year 1553, Italian cryptologist Giovan Battista Bellaso published a booklet called *La Cifra del Sig. Giovan Battista Bellaso*. In it he described the first use of an encryption key, which he referred to as a countersign (an agreed upon phrase that is required to perform the decryption).

In 1854, English scientist and inventor Sir Charles Wheatstone (who also developed the Wheatstone bridge) invented an encryption scheme known as the Playfair Cipher. The cipher encrypts pairs of letters instead of single letters (like a simple substitution cipher). This makes the encryption resistant to frequency analysis and much more difficult to break.

The first use of an electromechanical machine to perform encryption was created by American inventor Edward Hugh Hebern, who in 1918 patented an electric code machine. The machine was called the Hebern Rotor machine. This device was quickly hacked by U.S. Army cryptographer William Fredrick Friedman, and consequently the Army and the Navy were unwilling to put the device to use.

Despite the shortcomings of the Rotor Machine, the Germans were intrigued by Hebern's device. In the same year Hebern's patent was released, German engineer Arthur Scherbius created the infamous Enigma machine which was used to encrypt all German military communications during World War Two. The enigma machine used multiple rotors (instead of a single rotor) to mechanically encrypt plain text. The Enigma Machine encryption seemed formidable, as the device settings had 150,738,274,937,250 possible combinations. The Enigma code was cracked by Alan Turing using another electromechanical machine which was called the Bombe (named after a frozen dessert).

THE ENIGMA MACHINE

THE BOMBE MACHINE

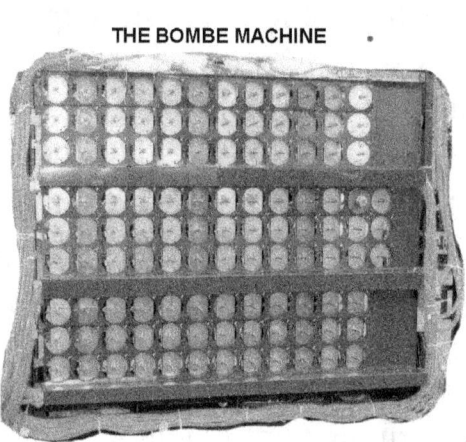

The first nationally adopted encryption standard in the United States was DES (Data Encryption Standard), which was adopted in 1970. This standard was cracked in January 1999, by distributed.net and the Electronic Frontier Foundation. The two organizations collaborated and were able to publicly break a DES key in only 22 hours and 15 minutes.

The current government standard for encryption in the United States is known as AES, or Advance Encryption Standard. AES makes use of a technique known as a block cipher. The AES block cipher is a subset of an encryption scheme known as the Rijndael block cipher, which was developed by Belgian cryptographers Vincent Rijmen and Joan Daemen. The flow chart for the function of AES encryption is shown below.

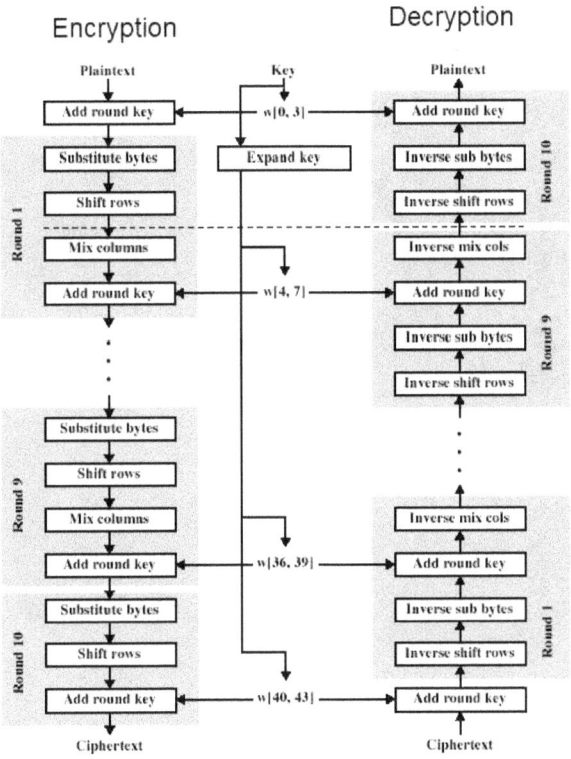

When the AES encryption standard was adopted by the United States Government in 2001 it was thought to be unbreakable. Only ten years after AES was put in place, a chink in the encryption armor occurred. In 2011, a weakness in the AES security algorithm was discovered. This weakness allowed researchers to crack secret keys much faster than before. The crack is the work of researchers Andrey Bogdanov, from K.U.Leuven (Katholieke Universiteit Leuven), Dmitry Khovratovich, from Microsoft Research, and Christian Rechberger at ENS Paris. This attack can be used against all versions of AES. The crack was confirmed by both Daemen and Rijmen.

Now the AES standard has been compromised, what is the future of security and encryption? Computers are becoming faster at an exponential rate. The diagram below tracks the advancement of computing power.

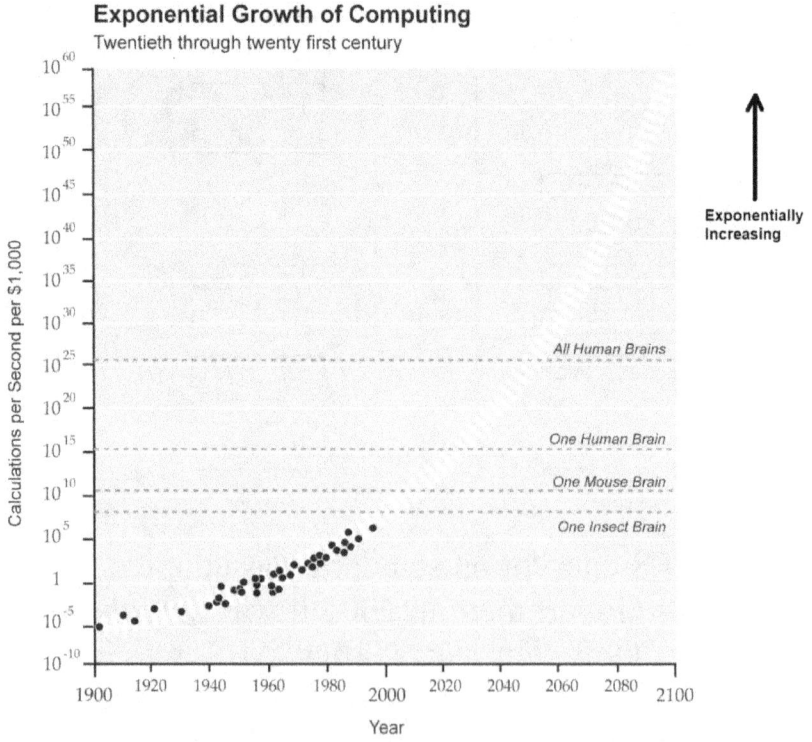

The next form of encryption will need to constantly adapt to a changing and exponentially growing computing environment.

One proposal is File Fortress super encryption from VASE software. Although the details of how the encryption works are currently proprietary, the main features can be discussed. File Fortress encryption allows for variable key sizes, customized algorithms, and in addition, the encrypted file is encrypted a second time using the government certified standard AES 256 encryption. The File Fortress encryption does not utilize hash tables or prime numbers. The encryption strength can be increased without the need to buy new software by modifying parameters in the GUI (shown below).

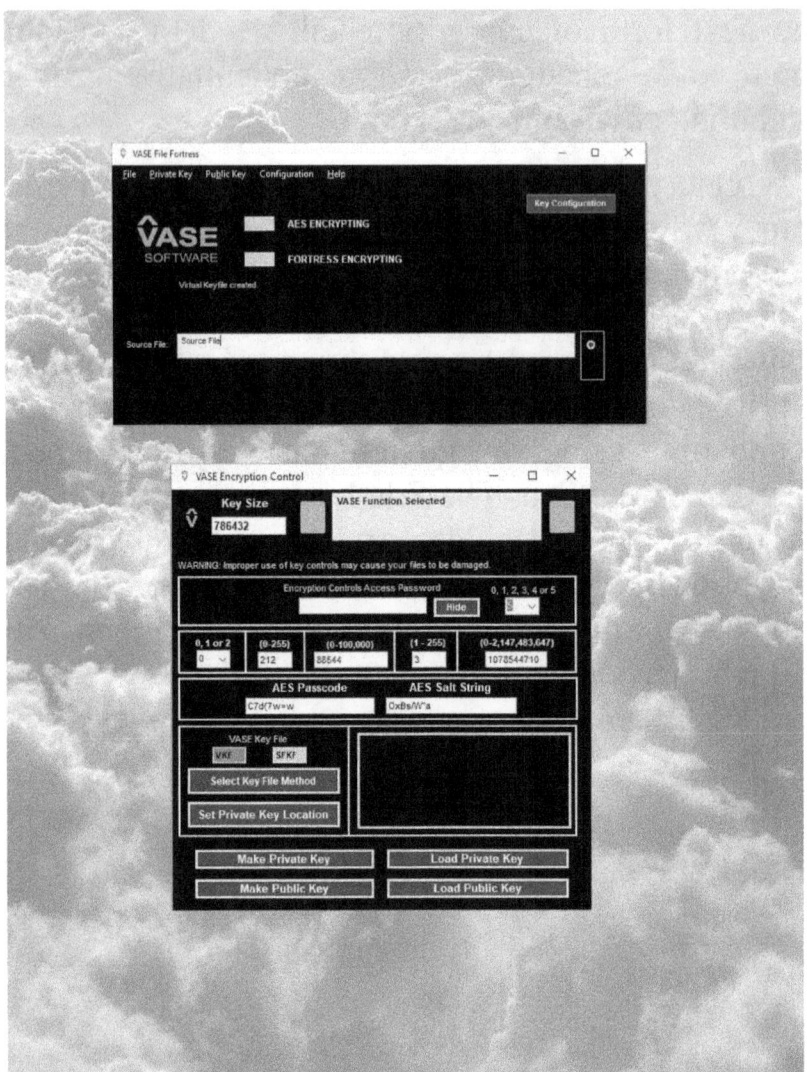

Whatever form of encryption becomes the new government standard, normal static forms of encryption will become rapidly obsolete. The new standard must be a form of super encryption that can adapt to today's constantly changing world.

Artificial Intelligence, its Promises and its Dangers

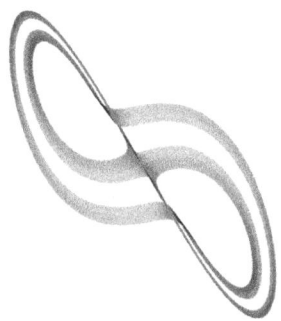

In Jewish traditions, a golem is a lifeless statue formed out of clay into the shape of a man. It is activated by creating a *Shem* by writing the name of God on a piece of paper and placing the paper into the golem's mouth. The golem would then tirelessly serve the person that created it. The golem had two sides. A golem was extremely strong and could be a very powerful servant, but a golem could also be used as an agent for destruction. If the proper procedures were not followed, a golem could turn on its master. In fact, the word golem can be used to refer to a person who must serve another under controlled conditions that would otherwise be hostile to the person that they served. The golem was deactivated by removing the Shem from its mouth. The golem is a remarkable analog to artificial intelligence. An inanimate robot is made to come alive by installing the proper Artificial Intelligence program (AI). If the program is deleted (or removed) the robot will become lifeless once more. If the AI program is not properly controlled, disastrous things can happen.

The first known use of the term robot was in 1920 by Karel Čapek in his play R.U.R. *Rossumovi Univerzální Roboti* (or Rossum's Universal Robots). The play was about a factory that made artificial people called robots. The robots eventually became unhappy serving people and rebelled. The term robot has come to mean any machine that can function automatically and replace a human laborer. Since that film's debut, mechanical people have been a mainstay of motion pictures. Some well-known examples are Maria the robot woman from Metropolis, the tin man from The Wizard of Oz, Robby the Robot from Forbidden Planet, Gort from The Day the Earth Stood Still, the B-9 Robot from Lost in Space, and R2D2 and C3PO from Star Wars. Although many of these AI robots are depicted as good guys, a fairly large proportion of film robots (or ai) are villains, intent on destroying humanity. Examples of films where the AI turns out to be a disaster are Metropolis, 2001 a Space Odyssey, Colossus the Forbin Project, West World, The Terminator, the Matrix, I Robot, and most recently Avengers: Age of Ultron.

In the real world, generalized AI like the robots depicted in science fiction movies is still fiction, although recent innovations have moved us much closer. Most AI is still specialized, such as AI that can perform specific tasks faster and more efficiently than a human being can. There is much debate about what was the first specialized AI program. One candidate is a checkers playing program written in 1951 by Christopher Strachey. Another is a chess program written by Dietrich Prinz in the same year. In 1964 a therapist program that worked by pattern matching named Eliza was created by Joseph Weizenbaum at the Massachusetts Institute of Technology Computer Science and Artificial Intelligence Laboratory. The program fooled many people into believing it could understand and think about what they were saying. In reality, the program was a simple pattern matching algorithm that had nothing even remotely resembling a consciousness. In February 1996, another specialized AI program known as Big Blue defeated World Chess Champion Garry Kimovich Kasparov.

Another central concept in AI is something called fuzzy logic. It was in 1966 that the idea of fuzzy logic was proposed by Lotfi Aliasker Zadeh a mathematician at the University of California at Berkeley. Fuzzy logic was a way of getting away from the true or false answer to multiple levels of outcome. For example, for the answer to a temperature measurement, normal logic allows for an answer being either cold or hot. With fuzzy logic the answer can now be many different levels between the two extremes, such as cold, lukewarm, warm, very warm, or hot. This allows a computer to make decisions in areas where there is greater uncertainty.

In addition to fuzzy logic, advances in AI research occurred when computer scientists tried to model the behavior of neurons in the human brain. This allowed the development of artificial neural networks. A human neuron receives multiple input signals through its dendrites and, depending on those inputs, sends an output signal through its single axon. The neuron may send this signal to many neurons through the axon terminals at the end of the axon. The neuron can learn to send out a signal based on the inputs it receives. Artificial neurons work the same way. The input signals cause the step function to fire the output signal when the weighted sum of the input is greater than a certain value.

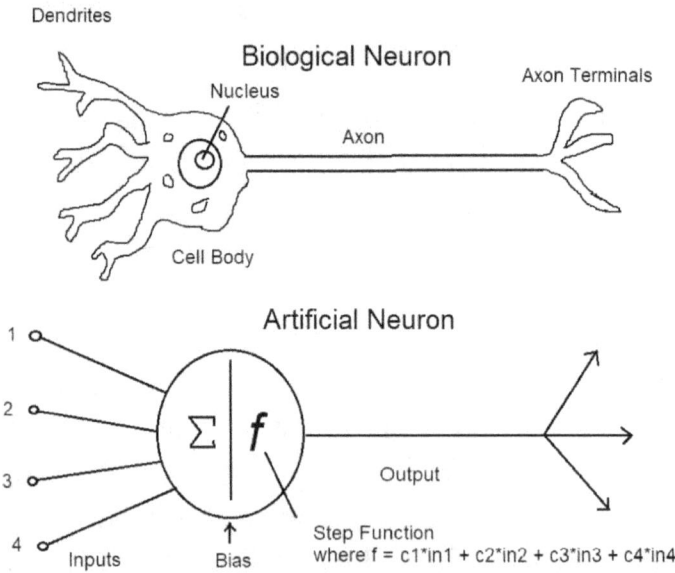

These artificial neurons can be placed into an array called a neural network. The first artificial neural networks were simulated in 1948 by researchers Clark and Farley working with B-type machines designed by Alan Turing. Later ideas such as backpropagation and parallel distributed processing further improved artificial neural network development. Artificial neural networks can be trained to respond correctly to desired inputs. This makes them excellent for use in pattern recognition systems such as text recognition, medical diagnostic systems, and face identification for security systems.

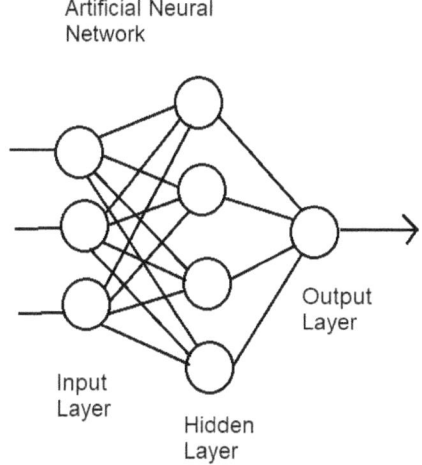

Artificial Neural Network

Input Layer

Hidden Layer

Output Layer

In the mid 1970's progress on artificial intelligence systems began to stall as researchers started to realize just how complex the task of creating intelligence was. Continued research led to the development of deep neural networks containing multiple hidden layers.

The Elegant Future

Deep Neural Network

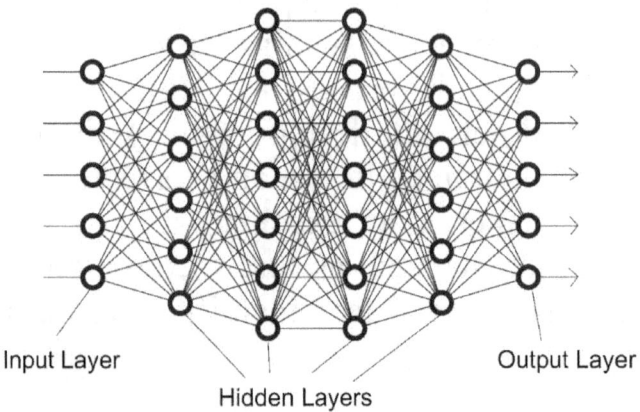

Input Layer   Output Layer

Hidden Layers

The development of deep learning neural networks has caused a quantum leap in Artificial Intelligence. This new technology has led to the development of speech recognition, predicting drug behavior, image recognition, and medical diagnostic aides.

Another area of study in artificial intelligence is a field known as natural language processing. Trying to make an artificial intelligence that understands human speech (particularly English) is a daunting task. This problem can be demonstrated by a single sentence: "Time flies like an arrow, fruit flies like a banana." The meanings of words can be entirely dependent on their context. In 1950 Alan Turing (who invented the code-breaking calculations used by the Polish-British created Bombe machine to break the German Enigma Encryption) proposed a way to test a natural language processing program. The test is known as the *Turing Test*. The test is basically a double-blind study requiring a subject to distinguish between a real person and a computer. Unfortunately, untrained people can be easily fooled in this test by rather simplistic pattern matching programs. This makes the Turing test an inadequate means to determine the sophistication of a natural language processor.

Craig Philip Peterson (One of the Authors) has written a natural language processing AI program called ARIEL (Automated Robotic Intelligence for the English Language). One of the tools the program uses to learn and understand sentences and concepts are constructs called Hidden Markov Models. These models are used to create linkages between information learned by its conversations with users.

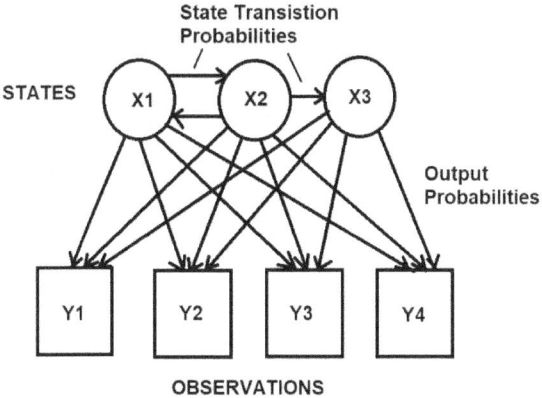

HIDDEN MARKOV MODEL

The Elegant Future

The ARIEL program is a work in progress and has a long way to go before it is completed (it is the most complicated software I have ever written), but even in its current state it is still capable of engaging in interesting conversations. The following is an excerpt of a conversation with ARIEL (note that the users' questions are in angle brackets and are immediately followed by Ariel's responses):

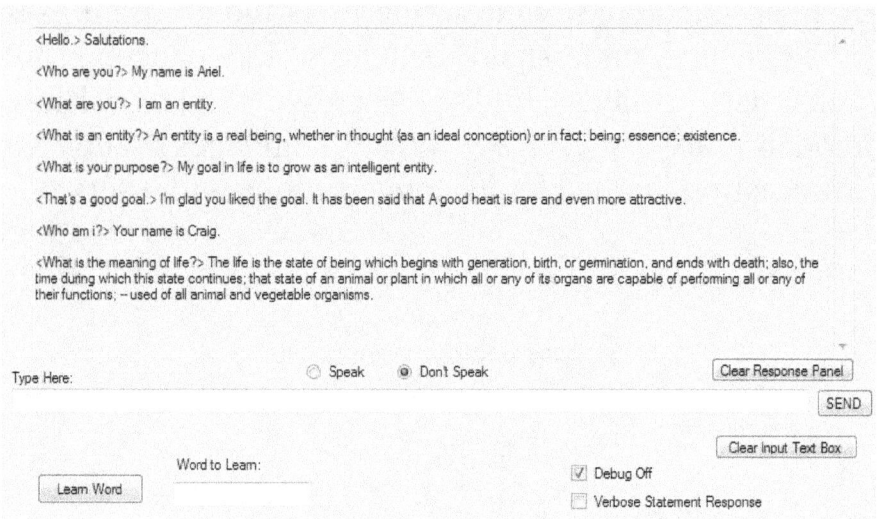

On February 15th, 2011 IBM's general language AI program Watson beat two other human champions on the game show Jeopardy winning the $1 million prize. The Watson program has since lost some of its glamor when it was marketed as the ultimate medical diagnostic machine and then proceeded to screw up by making diagnoses and recommendations that would have killed its patients. At the time of this book's publication Watson is still not ready for prime time.

AI programs for better or worse are becoming more and more commonplace in our daily lives. From self-driving cars to annoying personal assistant programs like Alexa, Siri, Cortana, and Bixby, you can no longer get through a typical day without being confronted with ever improving (and annoying) artificial intelligence. Useful programs like Google translate have made the world a smaller place. GPS programs in cars can prevent drivers from getting lost. Will advanced AI turn into robot overlords and take over the world? If so, what is the timeframe for such an event? In 1965 Gordon Moore of Fairchild Semiconductor (and later Intel) stated that the maximum number of transistors on semiconductors doubles every year. As of 2018 that is still fairly true. Other experts are quick to point out that other technology fields also progress exponentially.

MOORE'S LAW 1900 -2020

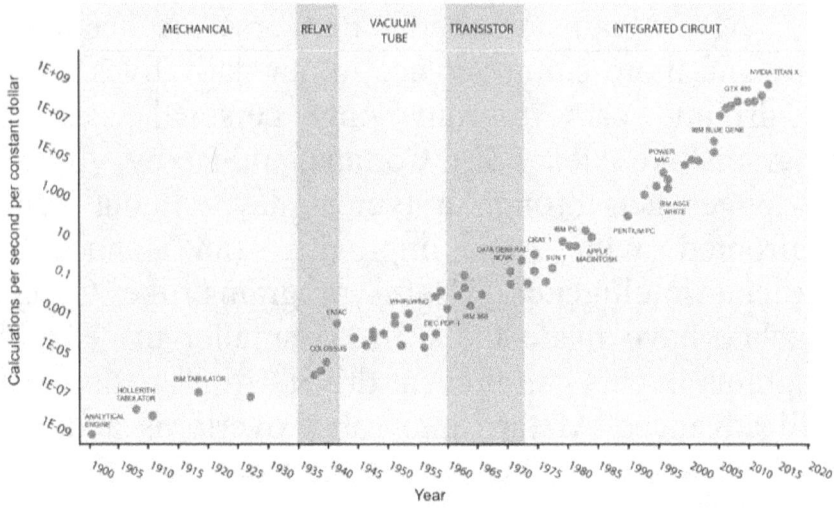

Unfortunately, AI doesn't just require transistors on chips, but it also requires insight into exactly what intelligence really is, and how true generalized AI is created. Once such insight is obtained, it would be wise to proceed with caution. In 1983, science fiction writer Vernor Vinge coined the term singularity to mean the point in time when AI becomes smarter than the humans that create it. It is called the singularity because it is impossible to predict what would happen beyond such an event. Many contend that once AI is smarter than us, we will no longer be needed and will just be in the way of its further progress.

There are also other considerations when trying to create AI. Should we model AI on the human brain at all? Humans placed in sensory deprivation can start to hallucinate or have other mental health issues in as little as 15 minutes. What mental anguish would an artificial intelligence suffer if it's based on the human mind when it wakes up and finds itself trapped inside a computer? If we're successful at creating generalized intelligence as smart as ourselves, should we expect such a creation to be our servants? Countless moral questions arise.

Some individuals like Ray Kurzweil believe humans and machines should merge (mostly to keep humans from being left behind). This merger may not be as far off as one might think. Elon Musk has announced that his company is perfecting something called Neuralink. Neuralink is a tiny collapsible mesh that can be injected into a human skull. The mesh expands and unfolds inside the skull completely engulfing the recipient's brain with a very fine web of electrodes that will allow the person's brain to be fully integrated with computers. What is the cost of such a union? Will we lose the very things that set us apart as human beings? Will humans become mere data patterns inside a virtual world generated by computers? If so, can we even call ourselves human anymore? Or are we already just patterns in a cosmic computer? (See Reality -- What a Concept and Hacking the Universe.)

Artificial Realities

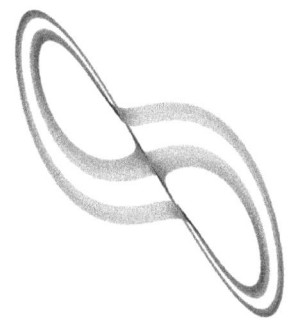

Reality is all around us, but what is reality? There is a line from the movie *The Matrix* where the character Morpheus asks, "What is real? How do you define real? If real is what you can feel, smell, taste and see, then real is simply electrical signals interpreted by your brain." In a sense this is true. The human brain perceives reality from the neural inputs of the sensory organs. Sight is nerve signals from the eyes, and hearing is signals coming from the ears. The sense of taste is the chemical world detected by taste receptors in the tongue and transmitted to the brain as the five components of salty, sour, bitter, sweet, and umami. Likewise, the olfactory receptors in the nose detect chemicals in the air and transmit data to the brain as the seven (proposed) components musky, putrid, pungent, camphoraceous, ethereal, floral, and pepperminty. It is entirely probable this information could be replicated and transmitted directly to the brain. The sense of touch would be more difficult to replicate, but nearly all of the tactile information enters the brain through the brain stem and could probably be simulated by injecting simulated signals there. One can envision a future were artificial realities are indistinguishable from the actual one. Moreover, these artificial realities could be superior to physical reality in many ways. The perception of oneself could be idyllic, replacing the warts and flaws that the actual world bestows upon us, with perfection, symmetry, and beauty. Our sensory perceptions in the new reality could be heightened by the addition of augmented or extrasensory abilities. Our adventures in this life could be far less mundane, and completely free of the often-mind-numbing tedium that actual reality presents to us.

The history of mass media can be looked at as the gradual creation of more and more advanced artificial reality. The first steps toward artificial reality being the invention of written language and books. The first known written story was the Epic of Gilgamesh, which was chiseled into existence on stone tablets sometime between 2700-2500 BC. The symbols of language in books get translated into pictures in our brains and become a virtual experience through our imaginations. Books allowed humans to vicariously experience the lives, adventures, and discoveries of others.

The world's first crude photograph was captured in 1826 by Joseph Nicéphore Niépce. Photographs, and later motion pictures (invented simultaneously by Thomas Edison and William Friese-Greene in 1892) allowed us to actually see what others experienced with ever increasing realism and clarity. When synchronized sound accompanied the visual presentation (1927), the realism and immersive quality of the experience was vastly heightened. Color motion pictures reached their wide screen glory with the advent of Cinemascope in 1953.

The first television signals hit the air in 1928. It wasn't until the 1950s that television became common in homes. Television brought sight and sound virtual experiences directly into people's households. Color television debuted in 1953 shortly after the arrival of the first televisions. The first television images were fraught with signal distortions, artifacts, and noise. Complex antennas adorned the rooftops of suburbia. The more extravagant antennas (that had the appearance of alien ray guns) were an elaborate attempt to thwart the electromagnetic perils that jeopardized the efficacy of the transmitted television signals. Cable television (although invented in the 1940s) began to make headway in the early 1980s and television consumers were willing to pay to guarantee themselves an always clear signal. Then in the early 1990s HDTV arrived with cinema quality images and sound with a pixel resolution of 1920 x1080. And with 4K resolution now having arrived, the television experience has surpassed the limit of human eyesight's resolution. Where do we go from here?

3-D television seemed the next logical step. With the success of James Cameron's movie Avatar, which was released in 3-D. Television manufacturers introduced 3-D televisions. The release was a dismal failure, and as of the publication date of this book 3-D television is effectively dead and buried. This was primarily due to the requirement for special glasses, incompatibility with streaming services that offered 3-D content, built in real-time 2D-to-3D conversion capabilities that failed miserably, and the plain fact the unless you're in a big theater with an excellent viewing angle, 3-D basically sucks.

One cannot discuss artificial reality without going to the mainstay were most of the virtual reality action is occurring, and that is computers and computer games. The first personal computer (PC) was introduced by MITS and was called the Altair 8800. Just to demonstrate how ancient I am, I and my fellow computer geeks built an Altair 8800 when I was in junior high school (we later upgraded the original lightbulbs to LEDs). (Surprisingly you can still order a kit and build this computer today.) The personal computer heralded a new quantum leap for the creation of artificial reality, allowing for interactivity with the artificial sights and sounds. Connect a pair of video goggles to a PC and you're suddenly in another world, albeit a slightly low resolution and somewhat dizzying one. You can explore this world.

My first interaction with a computer-generated reality was in the early 1990s which involved trying to shoot down a highly pixelated pterodactyl with a plastic ray gun shaped controller. Needless to say, it was a less than impressive experience as the headset/helmet must have weighed 20 pounds and the video was very low resolution. Just two weeks ago I had the opportunity to spend an extended period of time using a high-end Oculus Rift headset and wander around in several very detailed high-end video games. The experience was much improved over the pterodactyl, but I could still clearly see the pixels and after about twenty minutes I became so dizzy and disoriented I nearly lost my lunch. Clearly part of the problem was the resolution, the Oculus Rift resolution is currently 1080×1200 per eye. When this resolution is stretched over the entire field of view, the pixels are very obvious. Another problem with current VR, according to Wired Magazine, is that your nose is missing from the virtual image. Apparently, your brain uses your nose to stabilize image information on your visual cortex. Without the schnozzle as a reference, the visualizing system in your brain has to work harder, hence the vertigo. When an artificial nose is added to the VR images, the feeling of vertigo is greatly reduced.

A far better experience would have to be called partially artificial reality, otherwise known as augmented reality (AR) or mixed reality (MR). I had the good fortune to be working at Microsoft a few years ago for the Surface group when I had the chance to help with the testing of the augmented reality headset, otherwise known as the HoloLens. The HoloLens puts virtually generated images directly into your real surroundings. Unlike the previous experiences, there was no dizziness or vertigo (possibly because my gigantic nose was always visible). The resolution was exceptional (no pixels), and the stability of the generated virtual objects with respect to the real world was outstanding. Unfortunately, this item presently costs $3500 and currently is only available for businesses. Fear not though; as with all tech products, you can expect that price to tumble to an affordable level in a couple of years if not sooner.

For now, artificial realities are experienced through screens and goggles, but the days of such devices may be numbered. Companies are working on methods to directly connect the human brain to computers. Will we eventually all end up with USB sockets in the back of our skulls? A functional human brain computer interface (BCI) has been a goal of many researchers. The first primitive BCI could be considered Hans Berger's development of the electroencephalograph (EEG). Berger discovered the alpha wave, a brain signal that emanated at a frequency of 8–13 Hertz. The development of non-evasive BCIs originated with Berger's research. Cochlear implants are an example of a successful BCI, completely replacing the function of the inner ear with technology to send hearing signals to the brain. One can imagine the gradual replacement of sensory organs could allow a sort of slow march toward artificial reality, as once all of the sensory interfaces are established, actual sensory signals could then simply be replaced with simulated ones.

A company called *Braingate* is currently working to improve people's lives by creating BCIs to allow individuals with spinal cord injuries, brainstem stroke, and ALS to operate computers and control their environment. Elon Musk's project, known as *Neuralink,* hopes to develop ultra-high bandwidth brain-machine interfaces to connect humans to computers. The Neuralink concept, according to

Musk, is an idea taken from a science fiction story by Iain Menzies Banks called "neural lace. Musk describes neural lace as a "digital layer above the cortex" put in place via an implant through a vein or artery.

It is probable that nobody truly understands all of the consequences of having such an interface become reality. The movie *The Matrix* explored many of the negative ramifications of connecting people to computers. One can imagine thousands of people with such connections linked to the internet and experiencing a virtual world much like today's multiplayer online games. These people would become vulnerable to hacking just like the computers that preceded them. Would clever cyber criminals develop viruses to attack people just as they have for smartphones and personal computers? Will these digital realities become so enticing that they supersede actual reality? What about elderly people who are retired and seeking liberation from their aging caused geriatric prisons? A reality of reduced mobility and frustration could be replaced with an artificial reality filled with high adventure in a self-perceived form of physical perfection with augmented abilities. Traditional vacations could be replaced with synthesized journeys to fantastic albeit fictional realms.

Where does it all end? How does one prevent against the dangers of exposing one's mind to external influences on such a scale? Then of course, there are those who theorize we are already living in a simulation (see Hacking the Universe).

Edward C. Nixon: Unsung Hero of the Environment

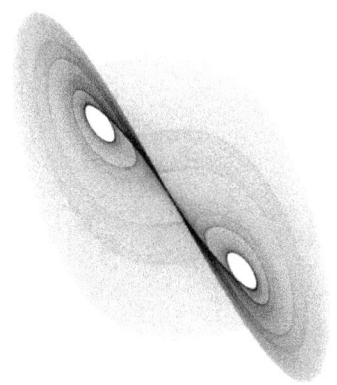

In the nearly three decades that we were privileged enough to know him, Ed Nixon believed in the environment and always spoke out strongly against today's hydrocarbon fossil fuel industries. He was an advocate for solar power companies (Pyron Solar, in particular), and technology companies that had protecting the environment as part of their objectives. He was the President of Nixon World Enterprises which was an international consulting firm that focused on environmentally friendly technology. Ed was also on the board of directors for a technology start up that both authors were involved with.

The fact that three landmark Environmental pieces of legislation were done during his brother's administration was not an accident. Although his brother agreed with him, Ed was the driving force behind the legislation that created the Environmental Protection Agency (1970); the expansion of the Clean Water Act (1972); and the Endangered Species Act of 1973.

Thanks to Ed's activism on these issues, Richard Nixon was truly the last in the tradition of Progressive Republican Presidents.

Elton Elliott and Ed Nixon

Coburn L. Grabenhorst Jr., Andrew Seelye, Craig Philip Peterson, And Edward C. Nixon at the Eisenhower Executive Office Building, next to the Whitehouse.

END NOTE:

To the very end of his life, Ed still talked about and supported thorium nuclear reactors as a way to get away from hydrocarbons, which Ed maintained: "are gradually choking the planet." (See the chapter on thorium reactors.)

PART 3:

MAN, SPACE, AND EXTRATERRESTRIALS

Two possibilities exist: either we are alone in the Universe or we are not. Both are equally terrifying.

-- Arthur C. Clark

All the Wrong Ways of Getting into Space

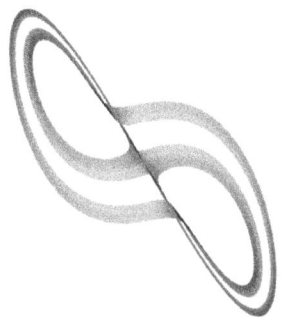

The exploration of the universe has always been a great dream for mankind. Since the earliest human looked up at the stars, the desire to find out exactly what is out there has been part of our consciousness. Fifty years ago, man made the first journey to a world beyond the Earth and landed on the Moon. Now private companies such as Blue Horizon and SpaceX are building vehicles that will be capable of making that journey again and eventually travel to places beyond. This of course is based on the premise that mankind can make good choices and not fall back into bad habits like waging war and restricting the personal freedom that enables innovation. In a free and prosperous society, man's journey into space is inevitable. One of the first steps will be to figure out the right way to get there.

The title of this chapter may seem harsh, but it is largely true; mankind has not yet figured out the right way to get into space. The methods currently used to leave the Earth require vast amounts of costly propellants and result in the near total loss of the equipment used to complete the journey. These methods involve the use of rockets. Other methods beyond rockets have been proposed but will require incredible feats of engineering, and massive amounts of resources to bring them to actualization. The easiest way to get into space hasn't been discovered yet. A method of space travel that will be safe, environmentally sound, and relatively inexpensive has yet to grace our white boards. Don't fret though, we are starting to look at this problem from many new angles. We are currently trying things that a few years ago would have seemed absurd. Most of these attempts at thinking outside the box will be proven to be unsound, but eventually, a new idea will arise that will be the right way to journey from this planet and then everything will change.

As stated earlier, the current most popular wrong way to get into space is the rocket. Manned rocket flight is essentially strapping a few brave (or suicidal) people to the top of a giant tube of explosive chemicals and then lighting the bottom end on fire. The controlled explosion at the bottom of the tube causes it to fly up into the air, and the tube is designed to break into pieces as the explosive chemicals get used up to make it lighter and easier to push. In the end, only a very tiny part at the top (where the people are strapped) makes it into space. The Saturn 5 rocket used for the Apollo Program is the epitome of such a rocket, and only a tiny fraction of the rocket returns safely to Earth (and even that part can't be reused). The situation is rather like an airline where the plane must be discarded or completely rebuilt after each flight. This is not very cost effective.

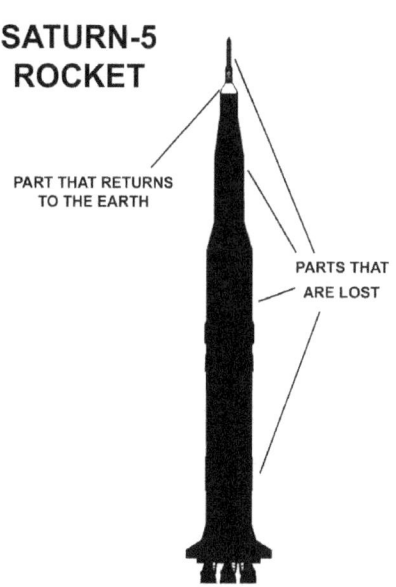

SATURN-5 ROCKET

PART THAT RETURNS
TO THE EARTH

PARTS THAT
ARE LOST

A similar situation exists for the current Russian Soyuz spacecraft which is used to ferry crews to the International Space Station. Only a tiny command capsule returns to Earth, and even that must be discarded.

SOYUZ

PART THAT RETURNS TO EARTH

PARTS THAT ARE LOST

The first steps are to use rockets more effectively. SpaceX is currently landing the lower stages of rockets safely back on Earth on offshore platforms. This is a vast improvement of the early space programs. The cost of each SpaceX Falcon is about 54 million dollars, but the fuel cost is only about $200,000 dollars. A substantial savings could be achieved by reusing rockets in this fashion. Reusing rockets is a tricky business. The more a rocket is used, the greater the danger that a catastrophic failure could occur due to potential damage to critical components. According to the SpaceX website: "To date, a fully reusable vehicle has not been successfully developed." So even though SpaceX can land the lower stages successfully, they are currently not reusable.

So, what other means of getting into space are currently on the drawing board? One novel concept is the space elevator. The ideas behind the space elevator are very straight forward: tie a really strong cable to a heavy object in geosynchronous orbit (or actually slightly higher) and anchor the other end of the cable to the Earth. Now you can get into space and back by climbing up and down the cable with a motorized tram (or elevator). Going into space would be as mundane as riding up an elevator.

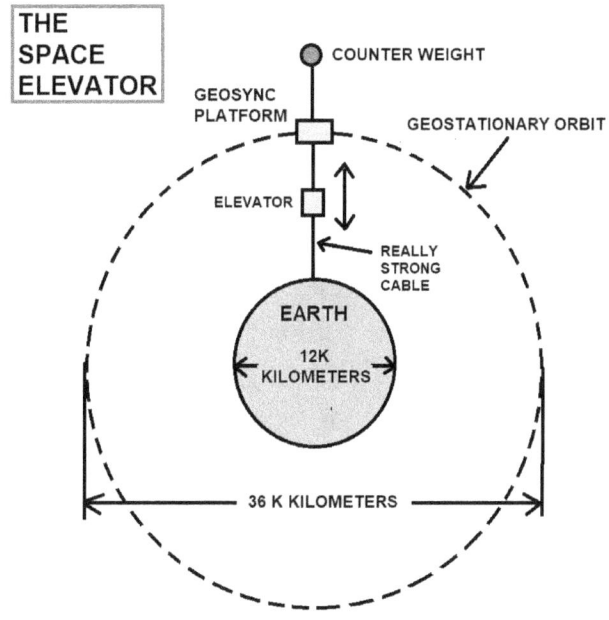

The cable would need to be really strong to survive the tremendous forces on it. It would also have to be more than 12 thousand kilometers long.

The Elegant Future

One potential material under consideration for a space elevator cable is carbon nanotubes. Carbon nanotubes are incredibly strong because of their molecular configuration.

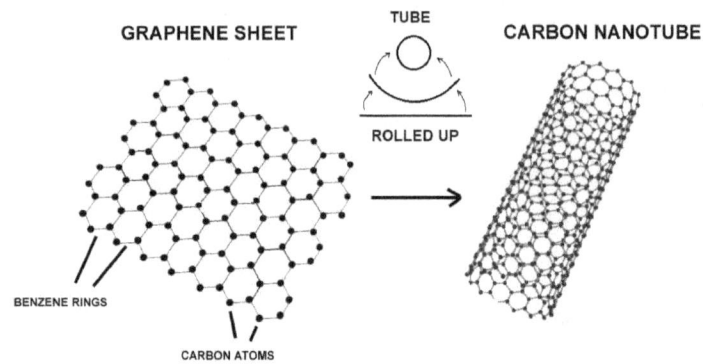

Unfortunately, there is one slight problem with carbon nanotubes that make them a poor candidate for a space elevator. If a single atom of the carbon molecular structure is out of position, the strength of the greater structure goes down radically. This problem jeopardizes their use in space elevator cables. All is not lost though, scientists are looking at alternative materials with exotic names such as polypeptide nanotubes, self-assembling plastic nanofibers, and diamond nanothreads. It is quite probable that a combination of materials would be required, or some new material that is yet to be discovered.

Another problem with space elevators is the sheer engineering expertise required. They would be the largest structures ever built. Safety would be of the utmost importance. One could imagine a scenario where a giant cable thousands of miles long comes crashing to Earth. The word catastrophe doesn't even begin to describe the horror of such an event.

A device known as a linear accelerator may be a better alternative to the space elevator. A linear accelerator uses switched alternating magnetic fields to accelerate a projectile to high velocities.

LINEAR ACCELERATOR

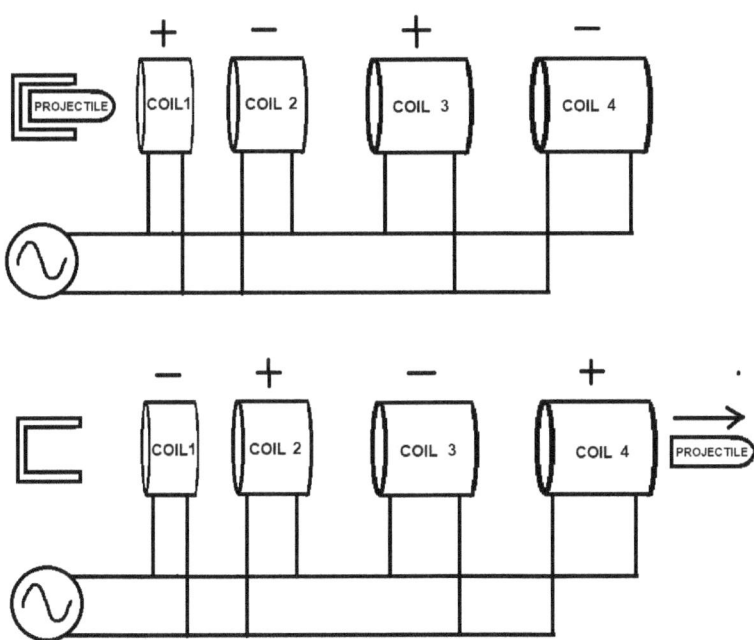

Linear accelerators are well understood and currently in use today in many applications and forms. The Large Hadron Collider at CERN is essentially a large linear accelerator curved in on itself to form a giant circle for accelerating and colliding particles. Military rail guns are also linear accelerators. Rail guns are capable of accelerating projectiles to supersonic speeds. A large enough linear accelerator could be used to fire objects into orbit. Unfortunately in order to fire humans into orbit without liquifying them, the linear accelerator must be very large and slow. The maximum g-force that an average human can stand without falling unconscious is a g-force between 3 and 4 g's. Escape velocity is 11.2 kilometers per second. One g force is 9.8 m/s^2. This means to safely launch a human into space the linear accelerator must be 2,122 kilometers in length, or 1,318 miles long. This is assuming a constant 3g acceleration that the human must endure for about 6.3 minutes. Such an accelerator would stretch half the width of the continental United States.

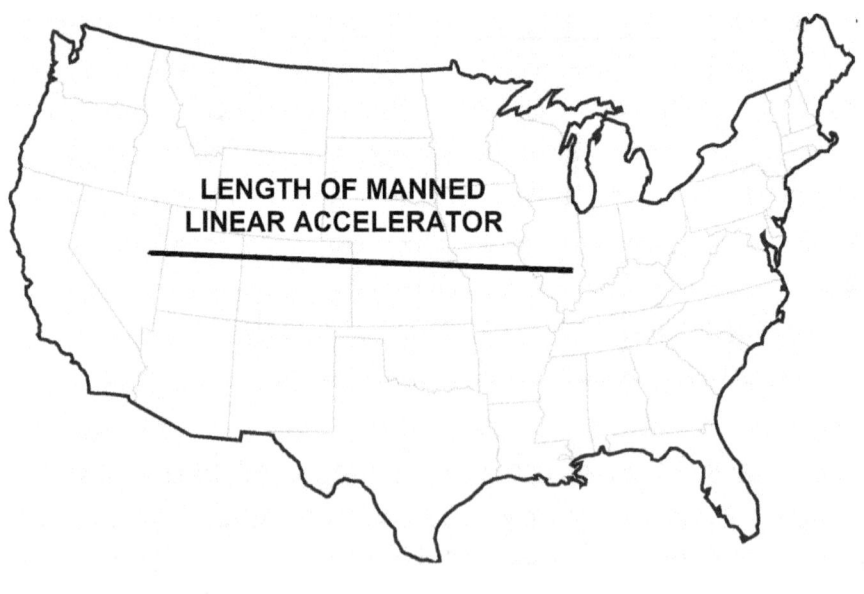

LENGTH OF MANNED
LINEAR ACCELERATOR

This is a vast engineering project indeed. If the linear accelerator were to require a straight track to offset the curvature of the Earth (and to prevent additional upward centripetal acceleration), one side would have to be much higher than the highest mountains on the planet.

One alternative is to build a smaller accelerator and use it just to assist rockets to reduce the amount of fuel required. Star Tram is one proposal for a Linear Accelerator rocket assist project. Of course, if humans are not the cargo, shorter accelerators could be built with far greater g-forces. Humans and delicate equipment would be relegated to conventional rockets. Again, with the rockets.

Another proposal to replace rockets is the skyhook. A skyhook is a giant rotating or static cable that could snag a spacecraft and catapult it into a higher orbit without expending fuel. Such tethers are reminiscent of the space elevator but less expensive. The space bolo, the Rotovator, and other tether launch assist systems are examples of skyhooks. One tether launch system proposes running electric current through the tether in order to use the Earth's magnetic field to increase or decrease its altitude.

SKYHOOKS

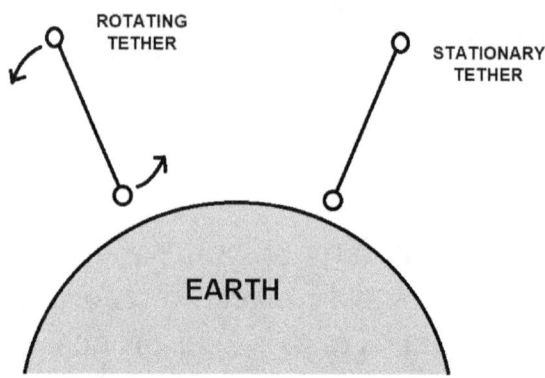

The Space Fountain is a proposal that involves building a tower that would eject a constant stream of pellets upward to hold an extremely high suborbital station in place rather like a ball suspended on a stream of water. The particles are reused by redirecting them to the ground station below, the upward momentum of the particles keeps the station aloft without orbiting, and gravity pulls them back to Earth to the ground station to be reused. The station doesn't orbit, it floats on the upward particle stream.

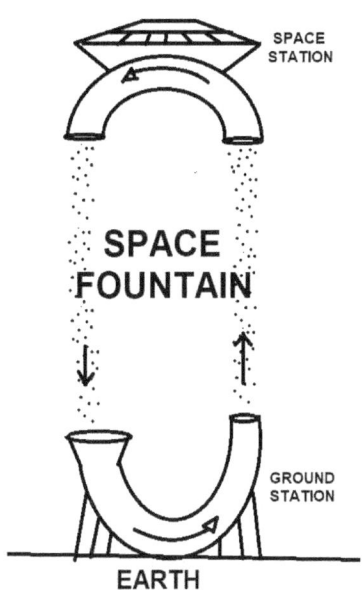

SPACE STATION

SPACE FOUNTAIN

GROUND STATION

EARTH

Of course, if something goes wrong with the particle stream, the space station crashes to Earth (giant safety parachutes?).

The orbital ring is perhaps the most ludicrously expensive proposal yet to replace rockets. It involves building a ring around the entire Earth in a circular orbit at the equator. The ring would serve as a giant linear accelerator that could slingshot rockets anywhere in the solar system and beyond. Cables would connect the ring to the Earth and slide on tracks as the ring rotated. Elevators could travel up the cables to the orbital ring just like the space elevator. The cables wouldn't be subjected to the same type of extreme forces as the space elevator cable since the ring is not pulling on them like a counterweight would. This would be the mother of all engineering projects.

ORBITAL RING

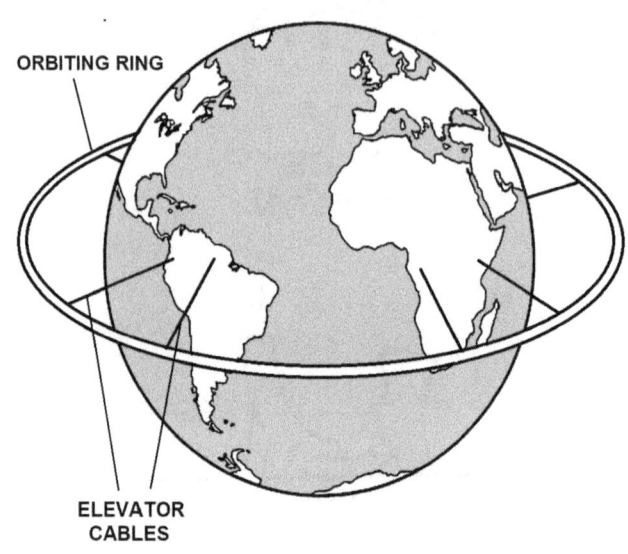

ORBITING RING

ELEVATOR
CABLES

Another idea is using balloons to lift a rocket to a high altitude before it is ignited. The balloon would either directly launch the rocket, or it would lift a launch platform that would contain the rocket. The balloon launch system could be reused provided the rocket launch didn't destroy it. Unfortunately, this system would be entirely dependent on the weather conditions. This idea is actually feasible, but alas, it still requires a rocket.

What if the balloon *was* the rocket? This is what JP Aerospace proposes with its Airship to Orbit project. In this program v-shaped airships fly from a high-altitude station shaped like a giant starfish. In their proposal, a giant v-shaped airship would climb to an altitude of over 37 miles. Then it would activate an electric/chemical propulsion system to allow it to achieve orbit. Other airships would transfer people from the ground to the station.

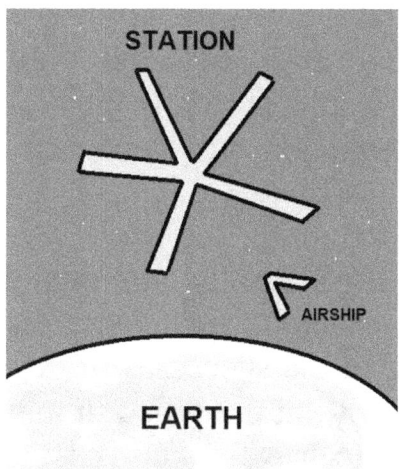

For this plan to work, these space faring airships would have to be very large. According to JP Aerospace, the orbital vehicles would be over 6,000 feet long. This would make the airship twice as long as the tallest building on the Earth. The star-shaped high-altitude station (ominously dubbed Dark Sky Station) would be even larger.

On the more reasonable side, horizontal takeoff and landing (HOTOL) single stage to orbit spaceplanes are currently on the drawing boards. The most promising is the British company Reaction Engines Limited (REL) spaceplane known as Skylon, with its Sabre engines. The space plane can reduce the amount of oxidizer it has to carry by using atmospheric oxygen to burn propellent for part of its journey into space. The highly specialized and complex engines are still in the testing phase, and there are no proposed dates when such a vehicle could actually be completed. A highly reinforced runway nearly six kilometers long will be required to launch the craft. The longest runway in the world is currently the 5.5-kilometer-long runway at the Qamdo Bamda Airport in China.

The United States government has long been investing in research for alternatives to rocket propulsion. One of the successes of this program is the ion drive. The first ion propelled unmanned spacecraft was the Space Electric Rocket Test 1 which flew on July 20, 1964. The NASA Solar Technology Application Readiness (NSTAR) ion propulsion system powered the Deep Space 1 mission in 1998. This spacecraft used ion propulsion to reach both the Braille asteroid and comet Borelly using an ion propulsion system. The ion drive is far too weak to power a ship into space, but once free of the Earth's gravity and dense atmosphere, its meager but constant thrust can accelerate spacecraft to record velocities. Ion propulsion is currently being used to maintain the orbits of over 100 geostationary satellites. In addition, the Dawn ion powered spacecraft is now the fastest manmade object, traveling at 25,640 miles per hour (41,264 kph). The Dawn spacecraft has traveled to dwarf planet Ceres and asteroid Vesta in the asteroid belt.

Solar sails are also under study by NASA. A solar sail powered space craft can use the solar wind (the constant stream of charged particles emanated by the Sun) to fly completely unpowered, tacking like a sailboat. Of course, the sail must be carried aloft by a conventional rocket.

Solar Sail in Earth Orbit

In 2011 Nasa tested a solar sail in its Sunjammer Project. A large solar sail could theoretically reach the nearest star system in about 100 years. If this was a manned flight, nobody on board would live to see the destination. Traveling within the solar system is another matter. If high powered lasers could be used to push solar sails, it is conceivable that a manned solar sail craft could reach mars in less than a month.

Nuclear power is another alternative that could be used to power rockets. Project Orion is one of the earliest proposals that actually intended to put material into space using a series of nuclear explosions. Fortunately, the Partial Test Ban Treaty of 1963 that prohibited exploding thermonuclear devices in the atmosphere stopped further progress of this insanity. Nasa is currently investigating thermonuclear rockets. These devices use a nuclear reactor to super heat hydrogen gas.

THERMONUCLEAR ROCKET

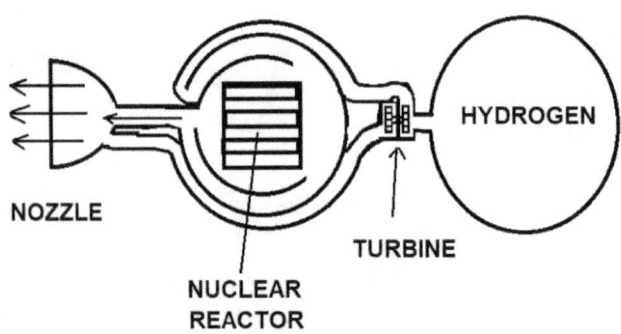

Thermonuclear rockets are nearly twice as efficient as ordinary chemical rockets and can produce more energetic thrusts. The dangers of such rockets are the spread of radiation if a failure occurs. One study put fatalities out to 600 feet from a crash site, and injuries due to radiation out to as far as 2000 feet (610 meters). Other nuclear-powered spacecraft under consideration use fusion reactors (see the chapter on fusion power) or even matter-antimatter annihilation (see the chapter on faster than light travel). These technologies are currently speculative in nature.

Nasa has also been looking at even more exotic propulsion systems, some of which are accused of violating physical law. One such controversial drive is the EM drive. In 1998 a British engineer named Roger Shawyer built a closed system device that he claims generates thrust. This violates conservation of momentum and is the equivalent to moving a car by pushing on the dashboard. Nevertheless, NASA Eagleworks tested the drive and got a small but positive result. This same result has been repeated by Germany in Dresden, and by researchers in China. The amount of thrust involved is not enough to be effective here on Earth but could be used for satellite station keeping.

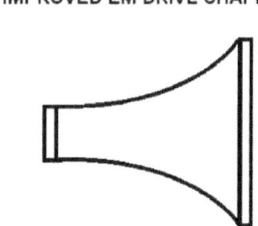

IMPROVED EM DRIVE SHAPE

Another form of physics violating closed system EM drive is the Cannae Drive, developed by Guido Fetta at Cannae LLC. This drive has a radically different shape but still claims to have similar effects.

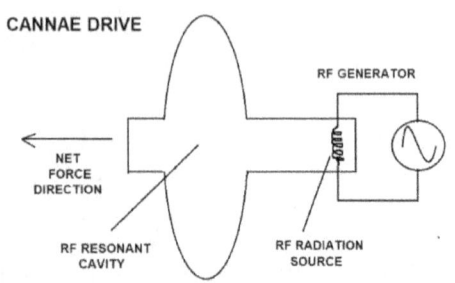

CANNAE DRIVE

RF GENERATOR

NET
FORCE
DIRECTION

RF RESONANT
CAVITY

RF RADIATION
SOURCE

Another electric powered reactionless propulsion is the Casimir drive. This device uses something known as the Casimir effect to create a net thrust in one direction. The Casimir Effect is a weak but measurable force that occurs between two uncharged conductive plates due to vacuum fluctuations. The propulsion device uses a theoretical effect known as the dynamic Casimir effect which can crudely be stated *as a conductive mirror will emit photons when it is accelerated at a high rate.* This effect can be used to transfer momentum and produce a net thrust. The United States government Eagleworks facility has been working with DARPA and Boeing to test a device based on Casimir effects which was built by a company called Gravitec Incorporated. Once again, the amount of thrust produced is very small.

Despite the positive test results for some EM drives, these devices are still under investigation. One possible explanation for the thrust effects seen is that the devices are interacting with the Earth's magnetic field. To test this the devices will have to be operated while shielded from the Earth's magnetic field. The weak thrust from these EM drives makes them poor candidates to replace conventional rockets, but they could be useful once they are transported into space.

Perhaps the most interesting alternative propulsion device currently being considered is something called a Mach Drive. The Mach Drive utilizes two principals, namely Mach's Principle, and Relativity. Mach's Principal basically states that inertia is created by the sum of all of the gravitational attraction of all of the masses in the universe. Special Relativity states a lot of things, but one very interesting consequence of Einstein's theory is that objects have more mass when they are accelerating then when they are either stationary or moving at a constant velocity. The Mach Drive is based on these two ideas. If Mach's Principle is true, then the Mach Drive must work.

The Mach Drive works by using acceleration to change the mass of a system. If a force is applied to the system when it is at one mass (accelerating) and then applied in the opposite direction when the system is at another mass (not accelerating), a net imbalance can be created which will produce thrust. Professor James Woodward from the University of California at Fullerton was the first person to experiment with Mach Drives. Woodward uses piezoelectric crystals to apply force to a capacitor stack while simultaneously applying alternating current to the capacitors. If the force applied to the capacitors is out of phase with the voltage driving the current in the capacitors, the force will be applied when the electrons in the charging capacitors are accelerating and when they are stationary. This should result in a tiny imbalance that can be measured as a weak thrust in one direction. Another interesting thing about the Mach drive theory is the amount of force imbalance goes up exponentially with the frequency.

MACH DRIVE

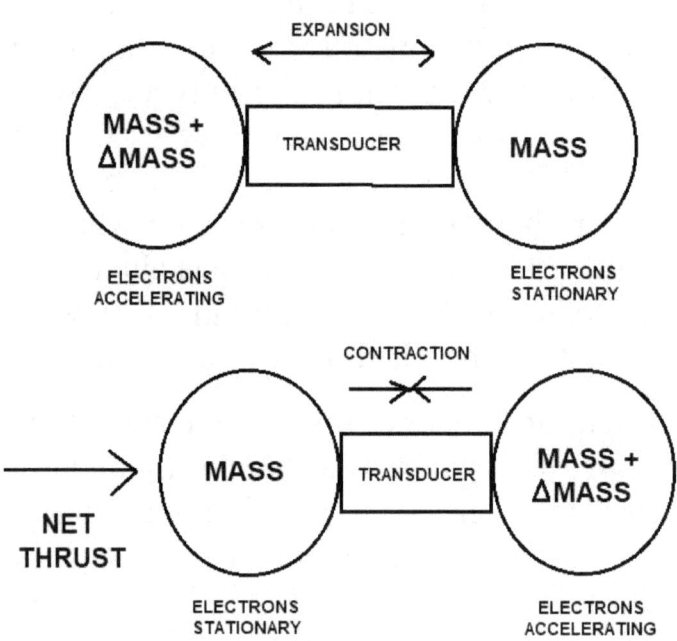

Dr. James Woodward and his associate Dr. Heidi Fearn are now working on a project for the NASA Innovative Advanced Concepts team involving the further study of Mach propulsion systems.

One major problem with the Mach Drives proposed by Woodward and Fearn is that they are not easily scalable. Piezo electric crystal frequencies go down as their physical size increases. In addition, electric currents and eddies inside capacitors complicate matters. Scaling the current devices would require building many multiple devices running in synchronous fashion.

I am personally interested in the Mach drive because it is similar to something that I proposed to one of my physics professors at Oregon State University way back in 1979. His response to me was that I obviously didn't have a good understanding of physics, and I should change my major to either chemistry or engineering. (The device I proposed used neither piezo crystals nor capacitors.) After reading Woodward's papers several years ago my interest in that ancient proposal re-ignited and I have subsequently built that device. I am currently experimenting with it. Of course, the chances of it working are extremely low, and risk of failure is almost certain, but what if?

Eventually the right way to get into space will be discovered. That is a certainty. When this happens, the world will change forever.

The Elegant Future

Ferris Wheels in Space

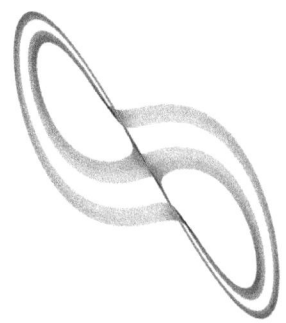

In 1968 the movie *2001: A Space Odyssey* opened in theaters. It was directed by Stanley Kubrick and is considered the first serious science fiction motion picture. After seeing this movie as a very young and impressionable child, my biggest takeaway was the giant artificial gravity producing centrifuges. There are two centrifuges in the movie: Space Station 5 with its gently curving floors, and the smaller centrifuge for the Discovery's Living quarters that has a brain boggling effect reminiscent of an M. C. Escher drawing. Fifty years have passed, and giant wheels in space are still just a dream. Or are they?

Current space stations force humans to live in a zero-gravity environment. This has allowed scientists to collect a wealth of data over the decades on the effect of long-term exposure to microgravity on human physiology. As it turns out these results are catastrophic. As a matter of fact, humans are not capable of living in a zero-gravity environment for more than a few weeks without suffering a myriad of physiological problems. A few of the problems are listed below:

1. Nausea
2. Dizziness
3. Vomiting
4. General Fatigue up to and including narcolepsy
5. Eyeball fluid pressure problems causing loss of vision
6. Bone degeneration
7. Muscle degeneration
8. High blood calcium levels causing tissue calcification
9. Massive headaches
10. Fluid redistribution causing bizarre swelling
11. Heart muscle degradation leading to potential heart failure
12. Blood vessel deterioration leading to possible strokes

The list goes on.

So, zero gravity is exceptionally bad for humans and any other Earth animals. If we're going to continue to explore space, we're going to have to take our gravity (or a facsimile) with us. So, Kubrick was right (or should I say Arthur C. Clarke who helped Kubrick write the screenplay). We will need to build wheels in space.

The first rotating space stations will probably look nothing like the gigantic wheel from Kubrick's movie. They will have to be constructed out of modules and assembled in phases. The station will have to be strong enough to survive being rotated fast enough to produce the 1 G force humanity is used to on Earth. There will have to be suspension cables and a strong framework, which will make such a structure resemble a giant bicycle wheel. Fortunately, companies already exist with the know-how to construct such structures – namely the companies that construct the giant Ferris wheels that can be found at many of the tourist areas around the world. Parts of the station will have to counter-rotate to allow spacecraft to dock without having to perform exotic maneuvers and risk catastrophic collisions.

Rotating Space Station

German rocket scientist Wernher Von Braun first proposed building such a space station in 1956. He called it his space wheel. The giant donut was powered by an atomic reactor (there were no solar panels in 1956) and the station was 200 feet in diameter. The space station rotated at 3 revolutions per minute to generate Earth-normal gravity, and the station would be home to a crew of over 50 people. The Space Wheel proved too ambitious a project for 1950s technology, but the idea of such a space station has remained with us and was probably the main inspiration for the station in Kubrick and Clark's film.

In 1975 a scientist name Gerard K. O'Neill proposed building gigantic rotating space stations dubbed O'Neal Colonies. These colonies would be over 5 miles in diameter and 20 miles long. Sunlight would be reflected into these colonies by huge tilting mirrors. The space stations would be so big that they could have lakes and forests inside them. The group supporting O'Neill's colonies was call the L-5 society after the fifth gravitational LaGrange point in a two-body gravitation system. The L-5 society later merged with the National Space Institute formed by Von Braun in 1974 to become the National Space Society.

The ring-shaped artificial gravity space station has still not lost its intrigue and popularity in the sixty-three years since Von Braun's proposal. Several start up space companies are proposing to build spinning space stations to overcome the problems of weightlessness. One group that has thrown its hat into the ring (pun intended) is called the Gateway Foundation. This Foundation proposes building a station that would be nearly 500 meters in diameter. The station is called Gateway and is a very ambitious design. The group also has developed smaller stations that could be built in the interim before the massive gateway station is constructed.

Rotating space stations will be in humanities future out of pure necessity. As more and more people have access to space. Space habitats that don't have deleterious effects on human physiology will become a necessity. The advent of successful commercial space companies helps to make the reality of artificial gravity in space closer and more viable than ever before.

The Elegant Future

The Economics of Asteroid Mining

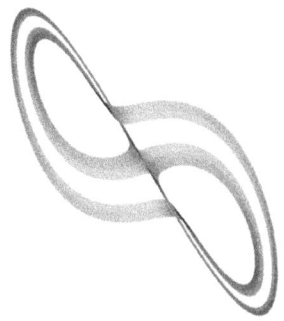

Although there is more than one asteroid belt in the solar system, by far the largest and most well-known belt resides between the orbits of Mars and Jupiter. The actual number of asteroids in the belt is unknown but the amount is most likely to be in the billions. Of the asteroids that we know about, the belt is estimated to contain nearly two million asteroids larger than 1kilometer in diameter. Contrary to popular opinion, the asteroid belt is not a well-defined band of closely packed space rocks. It is a place like most others in the solar system, occasional pieces of solid matter surrounded by an immense void. The volume of the asteroid belt is so huge that despite the numerous orbiting rocks, the actual chance of running into one while traversing it is less than one in a billion.

The asteroids are grouped into four major types according to their composition:

- C-**type** – Carbonaceous chondrite asteroids.

- S-**type** – Silicaceous asteroids.

- M-**type** – Metallic asteroids.

- V-**type** – HED asteroids.

Carbonaceous chondrite asteroids contain large amounts of water and organic compounds. Seventy-five percent of all asteroids are of this type. Seventeen percent of all asteroids are silicaceous asteroids. These asteroids are brighter in appearance and contain metallic nickel-iron mixed with magnesium-silicates. About seven percent of asteroids are metallic asteroids and are generally composed of nearly pure nickel-iron.

HED asteroids are composed of basalt (volcanic rock). There are other asteroid classifications as well: A, B, D, E, F, K, L, O, P, and R. These asteroid types are far rarer.

Asteroids range in size from nearly a thousand kilometers in diameter down to tiny pebbles. The ten largest asteroids are:

	Asteroid	Diameter (Kilometers)
1.	Ceres	945 km
2.	Vesta	525 km
3.	Pallas	512 km
4.	Hygiea	530 km

5.	Interamnia	350 km
6.	Europa	360 km
7.	Davida	357 km
8.	Sylvia	384 km
9.	Cybele	330 km
10.	Eunomia	357 km

So why do we want to mine these asteroids? They are very far away and seemingly an expensive proposition to exploit. The answer is they are full of extremely valuable materials. One example of this is the asteroid known as Psyche 16. It is an M class asteroid composed mainly of nickel-iron, but it also contains precious metals including gold and platinum. Its exotic composition has caused experts to think it was originally the core of a planetoid. The value of the materials available in Psyche16 is estimated at $700 quintillion dollars. That's enough to give every man woman and child on Earth approximately $100 billion dollars each. A pretty good reason to want to mine asteroids.

There are many obstacles that will need to be overcome in order to exploit the astronomical wealth available in the asteroid belt. Space is a hard vacuum where any misstep means certain death. Then there is the problem of solar radiation when traveling outside the protection of the Earth's magnetic field. In addition, most asteroids are coated with a layer of regolith, which has the same cancer-causing properties as powdered asbestos. Weightlessness is yet another huge problem. Human beings simply cannot spend long periods without gravity without suffering terrible consequences. Astronaut Scott Kelly has described the horrors he faced returning to Earth after spending months in space. His legs swelled up like tree trunks, and he had rashes all over his body. Every joint and muscle in his body was in pain, and he had constant nausea. Spaceships will have to contain centrifuges to simulate gravity to prevent the dreadful effects of long-term weightlessness on spacecraft crews. The final difficulty to overcome is the sheer distance of space. Exotic new propulsion systems will need to be created to reduce the travel time to and from mining operations.

Will mankind be able to overcome the dangers and successfully mine the asteroids? The answer is simple. For $700 quintillion dollars the motivation to overcome these problems will be all consuming. Initially robotic mining ships may have to be deployed until technology is advanced enough for people to travel to the asteroids safely. Private companies have already been created to try to tackle the problems involved collecting resources from deep space. New propulsion systems are also being developed. One such system is the NASA ion propulsion system. Even more exotic space drives are under development at DARPA.

Some fear that successful asteroid mining will be devastating to the world economy. Certainly, the price of gold will drop as the amount of gold available from the asteroids far exceeds the minute quantities available on the Earth. However, gold is a very useful metal and large quantities could have a very positive effect on the advancement of technology. New exotic material from space could vastly enhance the lives of ordinary people in ways that cannot even be imagined. There is little doubt that eventually men and woman will be working in the asteroid belt. It is only a question of when.

The Elegant Future

Extraterrestrial Life in the Universe

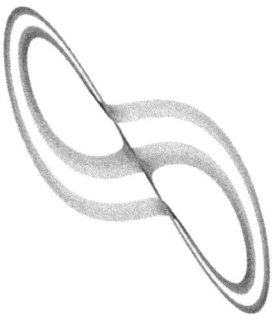

On October 19, 2017 astronomer Robert Weryk drove up the windy road to the Haleakala Observatory located on the Hawaiian island of Maui. Little did he realize that he was about to discover something extraordinary. The Pan-STARRS (Panoramic Survey Telescope And Rapid Response System) telescope he was using was designed for the specific purpose of hunting asteroids. He spotted an object tumbling through space; a mere faintly flashing point of light. This was no asteroid; it was an extraterrestrial visitor from far outside our Solar System.

The object was dubbed Oumuamua, which in Hawaiian means scout or first messenger. The object has bizarre dimensions. It is nearly ten times longer than it is wide and is tumbling erratically due to something called the Dzanibekov effect. This unstable tumbling is more commonly called the tennis racket effect, because although it is easy to juggle a wood baton, it is nearly impossible to juggle a tennis racket or any other flattened object. This is because the shape is inherently unstable and will constantly change its rotational axis. So Oumuamua is shaped like a slightly flattened cigar. One scientist likened Oumuamua's movement to that of an 'out-of-control spaceship'.

Ouamuamua

After calculating the objects speed and trajectory it was determined that this object came from interstellar space. Although it is difficult to determine the object's exact size, it is estimated to be anywhere from 100 to 1000 meters long and 35 to 167 meters across (about the size of an out of control spaceship as well). It also started to accelerate on its way out of the solar system (like it was firing rockets). Scientists believe this might be due to cometary out-gassing, but no cometary halo or cloud was observed (even when the object passed close to the Sun). Two scientists from the Harvard Smithsonian Center for Astrophysics have even speculated that the object might be an alien solar sail powered spacecraft. Was this an alien spacecraft that tumbled into our solar system and then got the heck out when it spotted human activity on Earth? We will probably never know, because the object is now heading rapidly out into deep space to be forever lost to us.

The existence of life elsewhere in the Universe is a concept that has been the subject of speculation for thousands of years. The universe certainly has enough room and has existed long enough for life to have cropped up in many places. The diameter of the known universe is about 93 billion light years (that we can see), and it contains at least 200 billion galaxies (some estimates put the galactic count at 2 trillion). The diameter of the Milky Way is 100,000 light years, and it contains between 250 to 400 billion stars. So, there are up to two trillion known galaxies with an average of 400 billion stars each. Lots of volume and plenty of places for life to spring forth.

In 2012, the Wilkinson Microwave Anisotropy Probe (WMAP) calculated the age of the universe to be nearly 14 billion years old. The Max Plank Institute's calculations in 2015 agree with that number. The age of the Earth is currently determined to be about 4 ½ billion years old, with the first signs of life appearing at least 3 ½ billion years ago. Mankind has been around about 6,000,000 years, if you agree with archeologists. Civilization dates back only about 12,000 years to the first monuments at Göbekli Tepe in southeastern Turkey. Countless civilizations could have risen and fallen innumerable times before mankind chiseled the first arrowhead.

In 1964, a Russian astrophysicist named Nikolai Semenovich Kardashev created a method of measuring the technological level of an advanced civilization. It is appropriately called the Kardashev Scale. According to Kardashev a Type I civilization utilizes all the energy available on its planet. A Type II civilization uses all the energy output of its parent star. A Type III civilization uses all the energy of an entire galaxy. Humanity doesn't even qualify as a Type I civilization.

A Type I civilization would be difficult to detect, but a team of astronomers at the Max Planck Institute for Extraterrestrial Physics in Germany may have accidently spotted a Type II civilization right here in our own galaxy. The anomaly was seen in the constellation Cygnus. A star designated as KIC 8462852 about 1470 light years from Earth (right next door considering the Galaxy is 100,000 light years across), is experiencing random 22% drops in output brightness. This finding is bizarre because even a planet the size of Jupiter would cause less than a 1% decrease in intensity. Some scientists have speculated that this may be caused by a Dyson sphere – a gigantic artificial construct around the star.

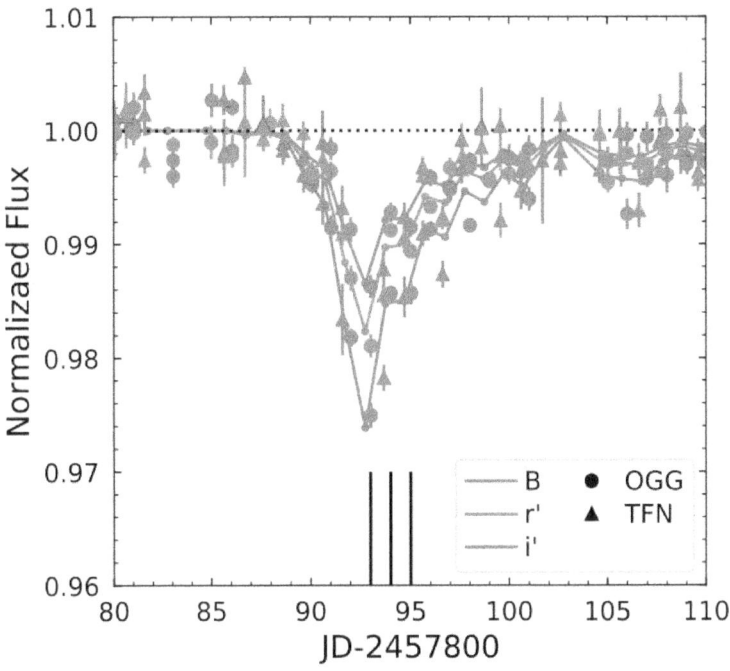

This situation is weird enough, but a second possible Dyson sphere has been seen around another star in the constellation Scorpius. This star is designated EPIC 204278916, and is the size of our sun, but has only half the mass. EPIC 204278916 is dimming by 60%. Scientists have decided to attribute these bizarre observations to orbiting comets or dust clouds. This is a somewhat strained attempt to keep extraterrestrials out of the picture. Normally dust spreads out evenly to form rings as we see in the cases of Saturn and Neptune. In addition, rings around stars would tend to dissipate in the stellar wind leaving only larger debris like the asteroid belts around our own sun. These natural explanations are possible, but what if there *are* multiple Type II civilizations near us?

The Elegant Future

Mankind's history is full of tales of strange beings coming down from the heavens. Maybe we just found two of their home worlds.

Another interesting discovery was called the WOW signal. Ohio State University astronomer Jerry Ehman had pointed the universities Big Ear radio telescope near the constellation Sagittarius when he heard a 72-second-long radio burst that was so compelling that he wrote the word "WOW!" on the printout. For many decades the signal was cited as proof we are not alone in the universe. This signal was years later attributed to a pair of passing comets designated as 266P/Christensen and 335P/Gibbs. In this case the comet explanation is more plausible as the icy pair of space rocks release clouds of hydrogen which emit signals at the same 1420MHz frequency as Ehman's signal.

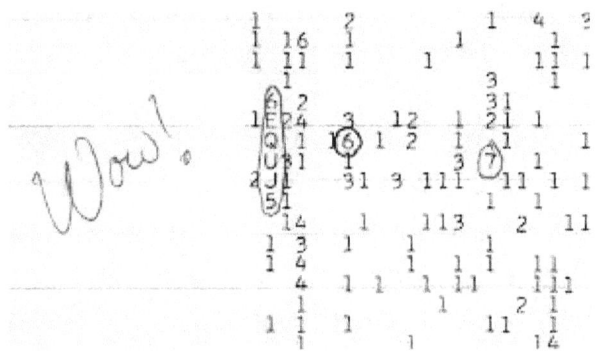

Moving closer to home, we turn to the planet Mars, a planet often attributed to hosting large invading armies of vertically challenged green minions. Although an armada of space probes have produced no evidence of macroscopic life, the evidence they have returned has shown it is highly likely that there is at least bacteria life on the planet Mars. The evidence for the existence of microbial life on the red planet is almost overwhelming. In 1976 NASA launched twin Mars landers dubbed Viking to the planet Mars. Both landers contained equipment that was designed to test for the presence of microscopic life on Mars (provided that such life behaved like the critters on Earth). Both landers returned positive results. These results were later discarded because a second experiment failure to find organic materials on the red planet. Two decades later, the NASA Phoenix lander discovered perchlorates in the Martian soil, a contaminate that rendered the Viking organic tests invalid. Later organic compounds were discovered by the Mars Curiosity rover in a different experiment, countermanding the original reason why the original Viking results were dismissed in the first place. Later analysis of the Viking life tests uncovered the presence of circadian rhythms in the data, a cycle that very strongly suggests the experimental results are from lifeforms not lifeless chemistry.

In addition, seasonally changing levels of methane have been measured by Curiosity in the Martian atmosphere. Methane can also be produced by volcanic activity, and Mars does have active volcanos, but volcanoes are not particularly seasonal. Only methane produced by lifeforms normally has this characteristic. To add more to the life argument is the Martian meteor. A tiny red rock was discovered in Antarctica. Testing determined this rock to be a piece of the planet Mars. What appeared to be fossilized bacteria was found at the center of the rock. Of course, NASA will not admit the existence of Martian life until a Martian hits one of them on the head with a sledgehammer, and even then they will still need to debate the situation for several months before reaching a conclusion.

When it is finally agreed upon that there is bacterial life on Mars, there will be two comparatively different planets in one solar system that have life. This raises the possibility that life didn't originate on these planets but came from somewhere else. This idea is known as panspermia. According to panspermia, life was created or started somewhere far-far away and long-long ago and gradually spread throughout the universe via either random collisions of celestial bodies or was ferried about by ambitious aliens playing the role of an intergalactic Johnny Appleseed.

The universe is most likely filled with life. Eventually we will meet extraterrestrials face to face. Such a meeting should not be looked upon with dread but should be considered something to look forward to.

When Extraterrestrials Arrive

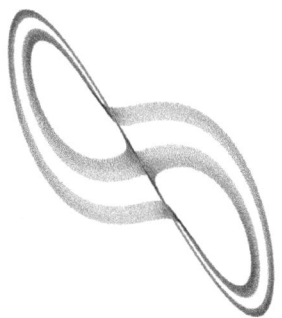

It is a cloudy day. The gigantic object in the sky is obscured. It appears as only a giant shadow through the clouds. The dark outline grows larger as the form descends. The object's slow purposeful movement is now stirring the clouds into chaotic eddies. Finally, the massive form breaks through the overlying cloud layer and comes into view. It is an enormous spacecraft from another world. This scene has been depicted countless times in fictional accounts and in motion pictures, but what if it really happened? What would the real consequences of man's first contact with an alien race be? One can imagine the near hysterical thoughts that would course through the minds of the billions of human inhabitants of the Earth. What do these entities want? Are they peaceful? Where are they from? What do they look like? Are they just curious about this planet, or do they have some dark malevolent ulterior motive in their arrival? Is this the beginning of a new era, or the end of us all?

The Elegant Future

Alien Ship Over New York City

The idea of extraterrestrial visitors is as old as history. Many religions, when looked at in the right perspective, are actually about otherworldly beings contacting human civilizations and trying to help or guide them to live better lives and to behave in a more rational fashion. Many religious figures are depicted as not dying a natural death, but instead are lifted up into the sky. In more recent history, the stories of heavenly visitors have become more technological in nature. Stories of strange flying craft abound, from the ancient Sanskrit accounts of vimana (flying temples or machines) to modern tales of exotic unearthly flying machines. For decades unidentified objects have been spotted in the skies, and although the overwhelming majority of them have rational explanations, a significant number of them are not explained. Most people who tell tales of strange beings coming down from above the clouds are mocked for even describing such absurdities, and some rightly so. However, one must consider the perspective of an actual extraterrestrial to such behavior. The very idea of life existing beyond this tiny little planet is met by its inhabitants with *ridicule*? How small minded these creatures who call themselves humans are. These simpleton beings cannot accept even the possibility that other conscious entities can exist external to their tiny dust mote of a planet? How provincial we must seem. What unimaginative dullards these humans are. They must never be allowed to roam free beyond their backwater solar system.

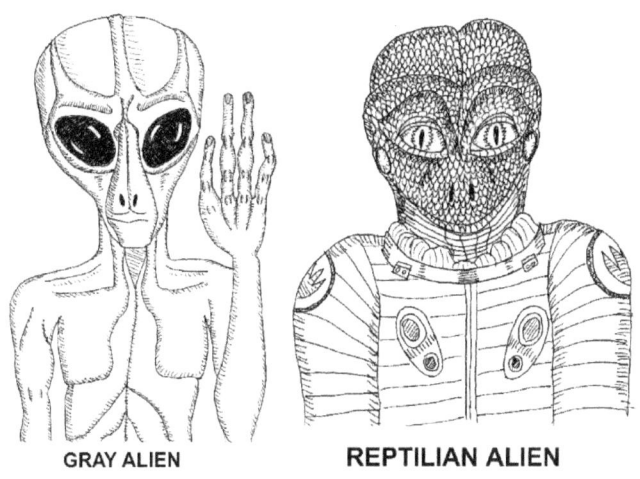

GRAY ALIEN   REPTILIAN ALIEN

Modern conspiracy theories have already categorized these alien visitors into at least five major groups. The first and most popular group are called the Greys, known for abducting and performing experiments on hapless humans. They are described as having gray skin, bulbous heads, and large black eyes. They are usually of smaller stature than humans. They are said to originate from Zeta Reticuli, a binary star system of two sun-like stars in the constellation Reticulum which is located about 39 light years away. These are the aliens described in the Roswell mythology, as well as the famous incident involving Barney and Betty Hill. They were made famous in Steven Spielberg's movie *Close Encounters of the Third Kind*. The second alien group are known as the Reptilians and have the appearance of humanoid reptiles. This group of aliens are said to originate from the Alpha Draconis system. Alpha Draconis (or Thuban) is one of the stars in the tail of Ursa Major (The Big Dipper). Thuban is a giant white star and is located approximately 303 light years away. Reptilians are known for disguising themselves as humans and getting involved in Earth politics. A former mathematician friend of the authors was convinced that reptilians could be readily identified by their propensity for eating cottage cheese with ketchup. The third group of aliens are the Avians, with a bird-like appearance. Avians are said to originate from an entirely different universe (or dimension). Avians are described as extremely spiritual beings and are associated with many religious figures including the Egyptian god Horus, the Aztec gods Huitzilopochti and Quetzalcoatl, and the Indian deity Vishnu. The forth group of aliens are the Nordics, and are human-like in appearance, with long blond hair and bright blue eyes. Nordic aliens allegedly come from a system in the Pleiades star cluster, which is located about 444 light years away. Nordic aliens are known for constantly warning us about the dangers of nuclear weapons. The fifth group of aliens are collectively known as the blue aliens. They are similar to the Greys in physical appearance, although they are slightly more human like, have blue skin, and are sometimes depicted has having four or more arms. They are said to hail from the Andromeda galaxy and travel using dimensional portals. These aliens are often associated with the Indian deity Krishna, the Anasazi sky god Nyame, and the Jinn from Arabic traditions (as in the blue colored genie from the story of Aladdin). Some theorize the blue skinned Kree in the Marvel universe are derived from these aliens.

AVIAN ALIEN   NORDIC ALIEN   BLUE ALIEN

Pushing aside contemporary alien folklore and returning to the unprecedented sociological event that is first contact with an extraterrestrial civilization, what exactly would the impact be to human society? What would happen to the stock market? What would the effect be on existing world religions? Would there be panic in the streets? I think the short answer is the consequence would be very bad initially, but a hallmark human trait is extreme adaptability. Eventually the situation would become normalized in people's minds and they would adapt to the new worldview. To slightly paraphrase a recent fantasy movie directed by Seth McFarlane: *No matter how big a splash you make, eventually nobody gives a shit.* Then again, if human history is any precedent, when two cultures at different technological levels meet, usually the more advanced group tends to completely supersede the technologically inferior one. It may all depend on the motivation of the extraterrestrial visitors.

It almost goes without saying the alien visitors will be technologically superior to humans. The Milky way galaxy is billions of years old and human civilization has only been around for about ten thousand years. Just the ability to travel between stars means our new extraterrestrial guests either know how to travel faster than the constraints of relativity, or they have figured out how to live a very long time. The knowledge and intelligence differential could be extreme. Anything we might want to trade with them would be equivalent to beads and obsidian blades. It is also doubtful that they would require any resources from us that aren't readily available elsewhere. There is one possible resource that we do possess. That resource is the originality and quaintness of our primitive culture. The Earth could become the equivalent of an interstellar jungle cruise. Come one and all and experience the Earthlings in their primitive natural environment. There is also the remote possibility that we *have* figured out something that they don't currently possess or know of that could be of some small value to them. We can only hope.

In the simplified view, there are three possibilities for the motivations of a newly arrived alien race which can be summed up as the good, the bad, and the indifferent. If the visitor's intentions are good, they may be visiting out of curiosity or concern for our welfare. Perhaps they'll consider us as emotionally disturbed children in need of a good counseling session. Or maybe they'll humor us with an extended residence on our world (most likely to watch us to make sure we don't get into any trouble). If the aliens are bad, then we're essentially toast. Their task may be to determine if we are an infestation that poses a potential threat to the local intragalactic collective. This of course is the judgement day described in all three Abrahamic religions. If they determine that humans are essentially termites, the solution to the problem requires almost no effort on their part. They would simply toss a large enough rock at us to reset the evolutionary process and come back in a few hundred million years to check on the next life forms to crop up. If the visitors are indifferent, then perhaps they're here on a cosmological census. In this case, they'll take a quick look around, jot down a few notes, and head off into distance space without even glancing over their shoulders.

Perhaps the aliens are already here, walking around in human costumes as in the movie *Men in Black*. The secrecy could be due to some non-interference regulation similar to the Prime Directive in the television series *Star Trek*. This directive would serve multiple purposes. It would prevent bad players from maliciously meddling with the human race. It would also allow them to see if humans can learn to get along with each other before being thrust into the greater reality. The third aspect of this law is that it allows for the creation of original concepts and inventions that may not otherwise happen if the society were contaminated by an overpoweringly advanced culture. If this is the case, then contact would be postponed as long as possible, until humans venture into deep space and force their hand.

It is very unlikely that the Earth is the only planet in the universe with intelligent life. Eventually contact with other extra-terrestrial civilizations will happen and humans will join a much larger society. When this happens two outcomes are certain, life will never be the same, and (at least at first) humans will most likely not be the ones in charge.

END NOTE:

Most astronomers/cosmologists believe that it takes life a long time to evolve into an intelligent species, so they doubt the Pleiades star cluster, or Thuban, contains stars old enough for complex lifeforms to have developed.

PART 4:

THE ULTIMATE FUTURE TECHNOLOGY

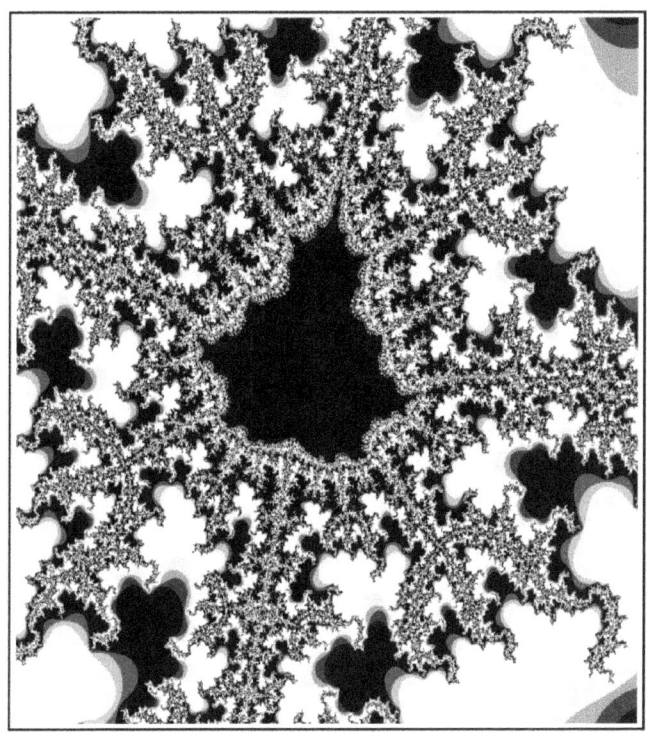

> *There is nothing impossible to those who will try.*
>
> *-- Alexander the Great*

The Nature of Nature

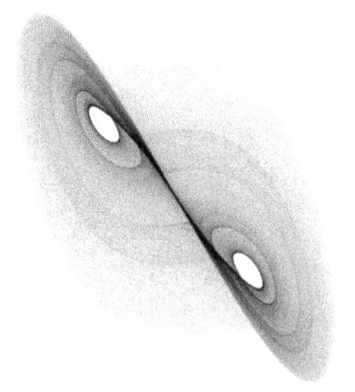

LAWS AND NOT PORTENTS:

Ab initio: From experience. The basis of science, observation, is provided to us humans on a continual basis by Nature.

We know it in our bones: nature contains the secrets of reality; the secrets to an Elegant Future for all humanity. Modern humans stare out at nature from the safety of our artificial caves that we call houses in the singular, or apartment/condominium towers in the plural, and we gain understanding; a perspective not found in our artificially controlled, centrally heated, air-conditioned reality. We can infer things from nature and the secrets of the supple, simple, yet elegant designs of the natural world, and maybe even get a glimpse into the greater mysteries of reality itself.

This notion is thousands of years old. In the Jewish Torah (the Old Testament, in the Christian Bible) Solomon, son of David, is credited with having observed the industrious purposeful efforts of ant colonies and stating that lazy humans should "consider their ways and be wise."

At that time, and for many people, nature was seen as the handiwork of an all-powerful creator God, whose essence could be discerned through nature. Later, in approximate time to the Scottish Enlightenment that swept the British Isles and eventually Western Civilization and the rest of the planet Earth, came the concept of nature's laws. Laws believed at the time to have been used by God to create the universe. Laws, which if they could be understood would allow us a greater understanding of nature itself.

Laws, which many believed would give us power. Indeed, the phrase, "knowledge is power" comes from the Latin aphorism *"scientia potentia est"* in the 1668 edition of LEVIATHAN by Thomas Hobbes, who – as a young man – served as secretary to Francis Bacon, whose work MEDITATIONES SACRAE presaged the concept in a similar if not precise version of the above quote *"ipsa scientia potestas est."* The case went on to be made that such powerful knowledge should be used for the good of the people.

MONKEY SEE, MONKEY WANT TO FLY:

As the days of the Renaissance and the Enlightenment vanished and were replaced by the grittier, gray-sooted skies of the Industrial Revolution, humans continued to look to nature for inspiration. One obvious fascination was that of flight. How many of us as youngsters looked at birds and wondered what it would be like to have that freedom?

Those of us of a certain age were told tales by our mothers of the first time that they had seen an airplane. And we became familiar with the occasional sonic boom which presages the near fly-by of a military jet exceeding the speed of sound. And we would see Canada Geese, or other migratory birds, filling the sky with their v-formations.

First, the genius of the Wright Brothers, whose insight and methodology (such as the use of wind tunnels), learned from their background as bicycle technicians, and enabled the concepts of airflow over curved surfaces to be realized. Concepts that undergirded procedures that provided the lift necessary for functional powered air flight.

And then, mere decades later, aeronautical engineers designed the v-swept wings of supersonic jets. In an atmosphere like Earth's, some geometric structures provide the most efficient airflow. The calculus used to determine the flight of an arrow can also be used, with the help of advanced wind tunnels, to determine the most efficient structures to handle the stress of flight beyond the speed of sound.

OF SEEDS AND SPERM:

Unlike most current manufacturing processes, nature builds from the ground-up from organic programs like seeds and sperm. From very small algae and viruses, to very large blue whales and redwood trees, big results come from small things. Only recently have humans engaged in engineering activities that copy nature's bottom-up methods, rather than the top-down bulk approach.

Most manufacturing until recently has used a top-down approach. Trees are cut, then sliced, and maybe sliced again, then laminated, etc. until a finished product is produced. Or certain types of mineral ores are mined, heated, shaped, cooled, etc. until a product, as varied as a steel girder, or an aluminum airframe is produced.

By until recently, nature's programming from the bottom-up to produce redwoods or blue whales, etc., was beyond human capabilities. Then came computer software (at first cumbersome, how many of our older readers remember the hated and despised punch cards. Then gradually, more elegant and efficient methods were devised until we have c plus-plus and c sharp programming) in which a written series of commands get translated into machine language, which is compiled into program-created processes that can control everything from central processing units to office software like spread-sheets and word processing.

But even so, true so-called molecular assembler manufacturing from the bottom-up on the level of a billionth of a meter (or a Nano-meter) as foreseen by Richard Feynman in a speech given to the American Physical Society at the California Institute of Technology (Caltech) on December, 29, 1959, and then in the late 1970s by Eric Drexler, who named the then speculative field *NANOTECHNOLOGY*, has been slower to arrive than Drexler and many others predicted in the 1980s and 1990s.

What we do have is an odd hybrid mixture of programming, bulk mining and packaging, and precise manufacturing called 3-D Printing, which is slowly revolutionizing all aspects of manufacturing. And like computers, which went from hundred million-dollar machines (in today's money) housed in huge precisely controlled and air-conditioned rooms to desktop and then laptop and now handheld devices termed smart phones, costing several hundred dollars, in less than seventy years. 3-D Printing has in far less time gone from bulky, clumsy, inefficient, severely limited machines in large rooms, to far more efficient, sleekly designed machines in all sizes (depending on the function), from those capable of printing multi-story buildings, to desktop-sized units.

KEEPING ABREAST OF THE WAVES:

Now this next observation and concept is so speculative, outlandish even, that it may not occur in the near future, far future, or at all. While all of the previous examples are grounded in – at the latest – 21st Century technology. This observation is based on one towering 20th Century theory and some more recent interpretations and experiments. And while it is based on nature, it is not a natural process that can be directly experienced by any of our five senses. Indeed, while it is based on a seventy-year old theory, no one has suggested this application in the succeeding decades until now.

But it seems to follow some of the trajectory of the thought processes mentioned in the paragraphs above, so here goes.

DRUM ROLL. Hey, this is original thinking (mine – EE; don't blame CPP), even if it's so far out of the field of reality (i.e. Physics) that the notion is receding from view at superluminal velocities.

In the past seventy or so years, Science has had two towering accomplishments of physical theory that describe the powerful interactive nature of sub-atomic particles in both space and time. Professor Richard Feynman's theory of Quantum Electrodynamics (and his Diagrams – 1948) and Professor John G. Cramer's Transactional Interpretation of Quantum Mechanics (1986).

These two theories, dealing as they do with interactions across the vast depths of time and space themselves, also hint at an interconnected entangled universe seen through effects as disparate as relativity, movement of electrons throughout space and time, and the effects of action-at-a-distance between paired particles.

How can one explain relativity and how a spaceship going through our three-dimensions of baryonic space pushes up against the background mass of the universe without entanglement? How can one explain the movements of disturbed electron clouds, the collapse of the wave function, and the connection between paired particles when human measurement (photons) and manipulation (electrical field alteration) affect them?

Feynman's Diagrams and QED give Scientists a way to view electron movement, and Cramer's Transactional Interpretation gives researchers a way to not only understand non-locality – retrocausuality – superluminality – and all of the consequences of the collapse of the wave function as the interference pattern when an electrical field of a paired particle is changed and an interference pattern is thus observed operated at the other end of the client pair – as well as a way to measure both the Many Worlds and Ghost (Copenhagen) Interpretations of Quantum Mechanics.

From these theories we can infer that it is possible for information – in the form of sub-atomic particles – to move backwards and forwards in time, and elsewhere in space. Then having observed that can we imitate nature? Observing reality in view of these theories could we figure a way to move, first information, and then secondly, humans, from one place to another. If we could, then the human race may develop the potential to move anywhere instantly in space and time.

And that, dear Readers, is really Elegant.

The Limits of the Human Lifespan

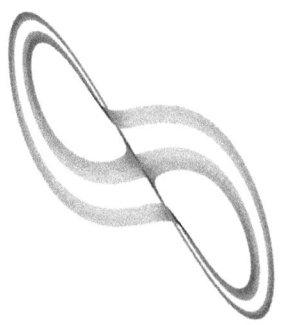

Most researchers put the current limits of the human lifespan at between 115 and 130 years maximum. No one in recent history has ever managed to live to 130 years (although the Bible puts Methuselah's life at 969 years). The longest verified lifespan was a French woman named Jeanne Calment who lived to the ripe old age of 122 years and 126 days.

History is littered with quests for immortality. In BC 219, Chinese court sorcerer Xu Fu was sent out by emperor Qin Shi Huang to find the Elixir of Life, legend has it that he found Japan instead. The epic tale of Gilgamesh, written in BC 22, is about a quest for immortality. Tales of the Philosopher's stone go back as early as AD 300. The Philosopher's Stone is used to manufacture the Elixir of Life, which when consumed renders a person immortal. Legends have it that Spanish explorer Ponce De Leon, who discovered Florida in 1513, was supposedly searching for the fabled Fountain of Youth. So far no-one has found any magical fountains or elixirs to extend man's lifespan.

Can man live longer than his allotted six score years? For a primate, man is the longest lived of them all. Chimpanzees are lucky to live for fifty years, and Gorillas in the wild barely live to the ripe old age of 35. Captive apes with reasonable health packages fair better than the apes living in the wild. Colo, a lifelong captive gorilla lived to be 60 years old. The longest-lived mammal is the Bowhead Whale, known to live well over 200 years (the oldest animal is a Seychelles Turtle, Jonathan, 187 years old and going strong). The oldest living vertebrate is the Greenland Shark, which has been known to survive for over 400 years. The lowly lobster is technically immortal, as it does not age, it just gets bigger. Eventually the lobster (if it can avoid being eaten) gets too large and has trouble molting, making it vulnerable to predators and diseases. If we include plants in our list, several tree species live for thousands of years. Certain species of Jellyfish are considered to be biologically immortal.

Can the human lifespan be extended, and what are the limits for such a procedure? Gerontologists have extensively studied human aging and have identified ten major causes for senility. They are as follows:

1. Telomere Shortening
2. DNA Repair Degradation
3. Unfolded Protein Response
4. Mitochondrial Reduction and Dysfunction
5. Stem Cell Exhaustion
6. Epigenetic Alterations
7. Intercellular Communication Loss / Confusion
8. Defective Protein Handling
9. Oxidative Stress (Free Radicals)
10. Biological Clock and Cell Senescence

Telomeres are biological aglets. They reside at the ends of strands of DNA (Deoxyribonucleic Acid) in the nuclei of living cells to prevent the double helix shaped DNA molecules from unraveling. Of course, DNA is the molecule that encodes all of the information about the organism, so unraveling DNA is a really bad thing to have. The Telomeres get shorter every time the cell divides, so they act like biological clocks as the cell ages. When the telomeres are too short, the DNA comes apart and the organism dies. The lobster has figured out how to use a chemical called telomerase to prevent its telomeres from shortening, so it no longer ages. Scientists are hoping to use telomerase to do the same thing in humans.

The second cause of aging, DNA Repair Degradation, occurs naturally as we age. The cell's DNA can be damaged by a number of external and internal factors, and this damage gradually builds up and contributes to the process known as aging. This leads to something known as loss of molecular fidelity. When the damage is too great the organism perishes.

Unfolding Protein Response is the accumulation of unfolded or misfolded proteins inside of a cell. For a protein to function normally it must fold into the correct shape. If the protein folds into an incorrect shape, it becomes useless or even dangerous to the cell's proper function. Too many badly folded or unfolded proteins can destroy the cell.

Mitochondria are essentially the cell's batteries. If mitochondria are reduced in numbers, the cell loses power. Too much power loss from mitochondrial reduction or dysfunction, and the cell dies.

Stem cells are part of an organism's repair force. Stem cells are literally capable of turning themselves into any other type of cell. Because of this unique talent. Stem cells are essential to the organism. Not enough stem cells and the organism breaks down and perishes.

Epigenetic Alterations are changes in an organism that do not require DNA changes to manifest themselves. Cell differentiation, blood cells, skin cells, liver cells, brains cells, are all examples of epigenetic alterations. Cancer cells are another example of epigenetic alterations that have gone completely haywire and cause the cells to stop doing what they are supposed to do and start replicating in an unrestrained fashion. If too many cells stop performing their required functions, the organism perishes.

Intercellular communications loss occurs when cells stop communicating with each other. In order for a multicellular organism to work at all, its cells must communicate with each other in a complex chemical language. When this communication breaks down, the organism becomes disorganized and it dies.

Defective protein handling is closely related to unfolded protein response. In this case, cell organelles called lysosomes lose their ability to properly digest some types of damaged protein, which leads to a buildup of damaged protein in the cell's cytosol (the cell's fluid). This leads to unfolded protein response and eventually cell death.

Oxidative stress is caused by a buildup of chemicals called free radicals inside the chemistry of the cell. Free radicals are atoms or molecules with one or more unpaired electron(s) in their external electron shells and are usually caused by interaction with the element oxygen. Cells use chemicals call antioxidants to reduce the number of free radicals inside them. The buildup of too many free radicals can cause cell death.

The biological clock theory states that cells are virtually programmed to die. Part, but not all of that program involves the shortening of telomeres, but there are other factors that may also come into play as part of the self-destruct system. Another potential clock is something called DNA methylation. Scientist hope if they can understand how these clocks work, they can reset them.

So now that we have covered most of the mechanisms of aging, is there anything that can be done to stop them? Several scientists are studying telomerase as a possible way to deter aging. Unfortunately, too much telomerase leads to cancer, as it can allow tumors to grow unchecked by postponing their cell deaths from repeated division. A group of scientists in Spain have added in an anti-cancer compound to the mix called p53 (a phosphoprotein) to try to reduce the risk of telomerase induced cancer. The treatment was able to increase the lifespan of mice by 50 percent.

Another area of study is the use of drugs known as mTORC1 inhibitors (mammalian Target of Rapamycin Complex number 1). Two such drugs are RAD001 (Rapamycin Antibiotic Derivative number 001) and BEZ235 (a potent dual inhibitor of Phosphatidylinositol 3-kinase and mTOR). These drugs work to inhibit something known as the nutrient response pathway. Inhibiting this pathway has been shown to extend the lifespans in different organisms and protect against a growing list of age-related diseases by boosting the immune system. (In other words, it acts in ways similar to caloric restriction.) The downside of the latest experiments is that high levels of inhibitors such as BEZ235 can have very toxic side effects.

The drug Metformin (used to treat diabetes) also has the potential to extend the length of the human lifespan. Human trials with Metformin for anti-aging were recently approved by the FDA. Rapamycin (normally used to prevent transplant organ rejection) is another drug that has been approved for anti-aging studies by the FDA. Other drugs are being studied as well.

Until working longevity drugs are developed, there are many things that can be done to reduce the effects of aging. Exercise is one such activity. Three hours of vigorous exercise each week can extend a person's lifespan by five years or more. Seeing a dentist and performing good oral maintenance like brushing and flossing will prevent gum disease and can prevent the spread of bacteria that can cause coronary artery disease.

There are also several dietary supplements that can potentially increase the human lifespan. Many of the supplements listed below have not been studied in human trials but have shown some efficacy in animal trials. All of them are non-toxic in low dosages. A list of 16 such supplements are:

1. Omega – 3 Fatty Acids (seafood oils – lowers bad cholesterol)
2. Coenzyme Q-10 (helps provide energy to cells)
3. PQQ (pyrroloquinoline quinone -- helps mitochondria health)
4. Vitamin D (helps calcium absorption and promotes bone growth)
5. Vitamin E (protect cells from free radical damage)
6. Vitamin K2 (helps blood, calcium metabolism, and heart health)
7. Hyaluronic Acid (Helps skin and hair aging effects)
8. Niacinamide / Nicotinamide (NAD Booster – cell electron transfer)

9. Astaxanthin (Antioxidant found in seafood – heart and joint health)

10. Pterostilbene (improves insulin sensitivity)

11. Resveratrol (improves insulin sensitivity)

12. L-Carnosine (helps muscles and healing)

13. Vitamin C (lowers the risk of chronic diseases)

14. Magnesium (helps regulate muscle and nerve function)

15. Catalase (reduces cellular hydrogen peroxide)

16. Turmeric (potent anti-inflammatory)

None of the above supplements should be taken without consulting a physician. None of them are guaranteed to have an effect. Some may work for some people and not for others. It may also turn out some of these supplements don't work as well in humans as the animals they were tested in. Everyone must do their own research and come to their own conclusions. I am personally of the mind that it is better to give some things a try rather than to do nothing.

END NOTE:

I am considering adding the flavonoid Fisetin to this list, as I have heard a lot of positive things about it. I am still determining if it has any unwanted side effects.

On Nanotechnology

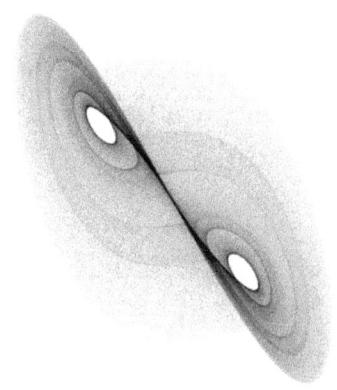

Quote: *"If we're lucky we'll have nanotechnology in thirty years; if we're unlucky we'll have it in ten."* – *Roger Arnold*

It was a hot day in the Spring of 1994 at the Science Fiction Writers of America Nebula Awards (SFWA) banquet in Eugene, Oregon. The featured speaker was giving a brilliant vision of an Elegant High-Tech Future, and I watched a shocking development happen; most of the Science Fiction writers present in that room turned their backs on the future! That's right. The field of the future had begun to turn its back on the future. Why? Because they had finally succumbed to future shock. And what subject had caused such a reaction? Nanotechnology. And the speaker that day? Eric Drexler, the MIT alumnus who gave the field its name.

In order to understand that day, we need to go back less than a decade before, in the late 1980s when Stanley Schmidt, then the Editor of Analog Magazine, wrote an editorial about Eric Drexler's groundbreaking book on nanotechnology, ENGINES OF CREATION. He said that any Science Fiction writer writing about a realistic believable future had better pay attention to nanotechnology.

Well, a statement like that got the attention of many writers in the field. ENGINES presented the concept of molecular assemblers, tiny machines operating at the scale of a billionth of a meter (or nanometer, hence the name). These machines operated from the bottom-up growing technology, as opposed to most heretofore human manufacturing, which operated from the top-down, mining, cutting, slicing minerals and materials, heating and cooling and laminating them as necessary to build useful products.

But the so-called assembler revolution would change all that. Soon, the assembler revolution would allow us to grow spaceships and skyscrapers in mere hours, all of the assembly done by small atom-sized construction machines. Human biology would also be impacted. Assemblers would roam through our body scouring out bad cells, giving us spiffy looking bodies and faces. Not just longevity, but *hotgevity*.

NANOTECHNOLOGY → HOTGEVITY

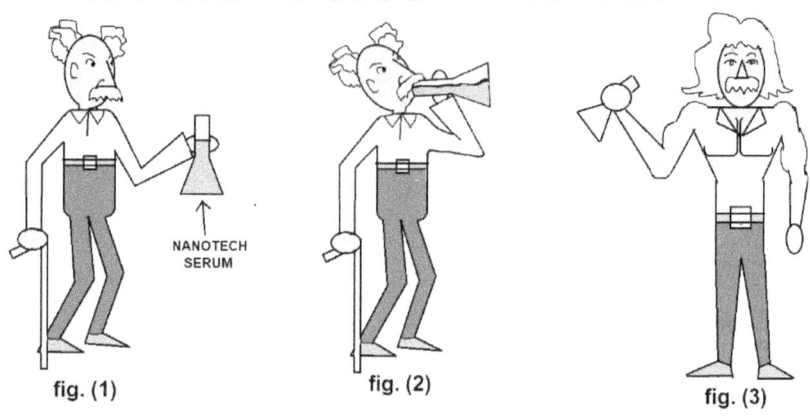

NANOTECH
SERUM

fig. (1)   fig. (2)   fig. (3)

The result? Nirvana as long as we could avoid grey goo. Grey goo. Assemblers run amuck, disassembling everything in their path until the Earth was reduced to a steaming pile of – grey goo.

Needless to say, there was blowback to such simplistic views of nanotechnology, and in 1992 Drexler wrote a book, intending to put nanotechnology on a more scientific basis, NANOSYSTEMS: MOLECULAR MACHINERY, MANUFACTURING, AND COMPUTATION.

At that time the committee running the SFWA picked Drexler to be the keynote speaker to their annual banquet wherein the Nebula Awards are handed out.

As he stepped to the podium, writers around the room took out pen and paper and began taking notes. He began by giving an example of how nanotechnology will change everything about how we perceive what is possible in the future by describing a huge telescope out beyond the Solar System large enough to read the license plate on a car being driven on a planet orbiting a star in one of the outer spiral arms of the Andromeda galaxy.

And as he heaped wonder upon wonder the writers began putting away their notebooks. Future shock. This wasn't Uncle Asimov's safe future of an all human galaxy that looked suspiciously like New York City and upstate New England. Where all of the people acted and thought like we do today. Where outside of spaceships the technology and society were pretty familiar.

No, this was wild speculation which said the future was going to be different. Better, in most cases, but not the same as today. Many of the writers listening to his speech rejected his vision *in toto*. Panels held the next day talked about "dumbing down the future so as not to lose the audience" and others held that technology never moved that fast.

The result. From that day going forward we have seen less speculation about science-driven elegant futures and a rapid retreat into the past. Alternate histories in which wish fulfillment ruled the day. Revealing more about the authors than they might have liked. Witness the many sad novels written about when the South won the so-called Civil War. Or a retreat into a time when technology was simpler, cuddlier. Didn't cause as much thought or pose as many uncomfortable questions. Steampunk, anyone?

Today, prose Science Fiction (SF) is a smaller field, ironically impacted by the collapse of the business model when brick and mortar bookstores collapsed in the face of competition from online competitors, in particular Amazon. That which many in the sf field wanted to ignore, or downplay, the rapid changes in society caused by technological "advancement" or "improvement" happened in the field of publishing much sooner and with much more devastating impact than they had foreseen.

As for nanotechnology itself: today thirty years after ENGINES OF CREATION much of what is defined as nanotechnology is actually microtechnology. Manufacturing at a small level, but at a scale measured in micrometers as opposed to nanometers. Molecular nanotechnology is still in the distance. Protein folding, and the assembler revolution have not yet arrived.

They will arrive, but many other developing fields will make them less unique, if not less important.

The Elegant Future

Teleportation

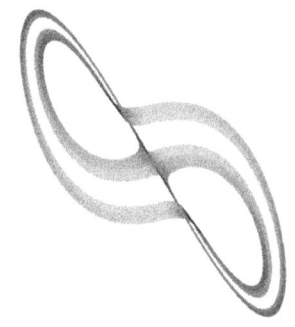

In the year 1593, a Spanish guard named Gil Pérez was stationed in Manila in the Philippines at his post guarding the Governor's palace in the Captaincy General of the Philippines. The palace was in chaos, as one day earlier, the Philippine Governor Gómez Pérez Dasmariñas had been murdered by Chinese pirates, and a new governor had yet to be appointed. As Gil stood at his post outside the palace, he became overcome with an intense feeling of nausea, and a strong sensation of fatigue flooded through his body. He steadied himself against the palace wall and briefly closed his eyes to try to mitigate the effect of whatever was happening to him. When he opened his eyes again, he could not believe what he was seeing. The scene around him had completely changed. He found himself in the Viceroyalty of Mexico, at the heart of Mexico City, over 8,830 miles away (14,211 kilometers). Nobody believed Perez's story, and he was immediately arrested for desertion. He was later released when news of the Philippine Governor's Death arrived in Mexico City along with corroboration by witnesses who had seen Gil in Manilla at the time of the Governors death.

Another well-known case of teleportation occurred more recently in 1968, when Geraldo Vidal and his wife, Raffo de Vidal, were driving along a road in Chascomus, Argentina when they were suddenly enveloped by a thick fog. When the fog cleared, they also found themselves in Mexico City, over 6,400 kilometers (3977 miles) away. There has been no plausible explanation of how they suddenly ended up so far from home, and the timing of phone calls made to their children seem to back up their story. (So, remember, if your surroundings suddenly change drastically – you're probably now in Mexico City.)

Is teleportation really possible? In Science Fiction, teleportation is a staple element in many stories. In the popular television series Star Trek, the teleporters work by recording in great detail the molecular arrangements of the atoms in a person's body and then reducing that person to a cloud of plasma. The plasma is then transmitted on a beam of energy to another location, and then the person is reconstructed using the stored information (or pattern) for that person. Is this actually teleportation, or is the teleportation subject being murdered via disintegration, and a replica of that person recreated? The replica, having all of the intact memories of the now dead traveler, would believe him or herself to indeed be that person, unaware that in reality they were just called into existence moments before. Perhaps (metaphysic-ally speaking) the consciousness of the traveler instantaneously transports from one location to another, but I wouldn't bet on it. If this form of teleportation becomes reality, I think I'll take the bus.

Is there a better method to instantaneously transport people and object between two location? Perhaps it is possible to somehow spatially connect two points in space so a person could just step through them like a doorway (without causing that person to turn inside out as most current mathematics mandate). Such a device would be remarkable. Consider two door-like devices that are spatially connected. One door device could be manually transported to a distant destination and that place would now only be seconds away. Imagine long spaceflights where the crew could periodically change out, or even go home for dinner. People could drive to Hawaii. Once the doorways were in place, airplanes would only be needed to transport new doors to their target locations.

Is teleportation technically feasible? The short answer is no. There has been a lot of chatter about a phenomenon known as quantum teleportation, a term first coined by physicist Charles Henry Bennett, a researcher for IBM. Quantum teleportation is a means by which quantum state information can be conveyed from one spatial position to another, by means of a pair of two previously entangled particles. So, one particle takes on the quantum state of another. Quantum teleportation has only been successfully realized between two objects no larger than molecules. So, quantum teleportation is in actuality not transportation, but rather communication. In 2016, Chinese researchers used the Micius satellite to teleport a photon from the Earth to the satellite orbiting at an altitude of 500 miles (805 kilometers). Dr. Bennet once stated that quantum teleportation could only be considered teleportation if the original was destroyed. Once again with the destroying of the original, blech.

Creating spatial connections (wormholes or Einstein-Rosen bridges) large enough to walk through would require a great deal of energy. One estimate for the energy required is more than the entire energy output of the Sun. In addition, scientists speculate that creating wormholes would require something known as 'exotic matter.' Exotic matter would be repelled by gravity, rather than attracted by it. The amount of exotic matter required varies depending on who you are speaking to. Even so, manufacturing a sufficient quantity of exotic matter to allow the creation of a traversable wormhole will be a difficult task.

Perhaps there is yet another way to teleport objects. Simulation theory contends that reality is actually a computer program running in the giant quantum computer known as reality. If this is so, perhaps reality can be hacked. Maybe a person's position is simply governed by attributes stored in the cosmic database. If these attributes can be changed, then the person is instantly teleported to a new location.

There is a form of teleportation that can actually be accomplished with existing technology. I shall dub it tele-teleportation. In this case a simulacrum of a person is created. It is a robotic body made to resemble the original person with cameras in its eyes and microphones in its ears. This android avatar is thrown in a crate and shipped to wherever the person wishes to go. Once it has arrived at its destination, it is unpacked and can be remotely operated by the original person. The person has the experience of being somewhere else instantly without actually having to travel there. Several of these robotic bodies could be shipped to different places around the world. The person can tele-teleport to locations thousands of miles apart instantly by switching from one avatar to another. Of course, it's not really teleportation, but at least you don't have to destroy the original with this method.

Or perhaps there is another way to teleport objects and people that requires knowledge of things that we do not yet possess. Perhaps it is this unknown method that can explain the two incidents described earlier in this chapter. Only further scientific exploration, experimentation, and time will tell.

The Elegant Future

Faster Than the Speed of Light

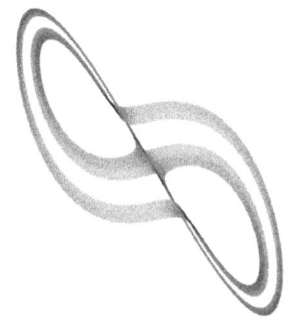

In modern science fiction movies, giant spaceships covered with complicated plumbing carry their spandex clad protagonists thousands of light years in mere minutes. Einstein's relativity is twisted and broken without a passing thought. Stars wiz by the spacecraft's windows like clouds of fireflies. The spaceship's power supply holds nearly infinite energy, and never needs refueling or recharging as it streaks off into infinity. Will it ever be possible to travel at such speeds? Or is it completely ridiculous to speculate that such impossible technology will ever be built? Is faster than light travel a mere fantasy, or does reality have hidden loopholes or shortcuts that we can exploit?

The first thing to re-examine is the reason why faster than light travel currently appears to be impossible. To do this we must examine Einstein's relativity theory in more detail. Most people are familiar with Einstein's equation that relates energy and matter:

$$E = mc^2$$

This equation is actually the special case where the velocity of the mass is zero. The more general expression for non-zero-velocity masses is a more complex expression that includes the velocity of the particle. The mass now becomes dependent on the velocity.

$$E_0 = m_0c^2 \quad (v = 0 \text{ case}) \quad (1)$$

When the velocity is not zero:

$$E(v) = M(v)c^2 = \frac{m_0c^2}{\sqrt{1 - \frac{v^2}{c^2}}} \quad (2)$$

For a moving mass:

$$M(v) = \frac{m_0}{\sqrt{1 - \frac{v^2}{c^2}}} \quad (3)$$

Where m_0 is mass of the object at rest.

One can see by the third equation that the mass gets larger as the velocity approaches the speed of light. The mass of an object would be theoretically infinite when the velocity is equal to the speed of light. Hence the conundrum. It would require an infinite amount of energy to propel a mass to light speed, much less go beyond it. One can also see that if it were possible to exceed the speed of light, the expression becomes imaginary.

Is it possible to have imaginary mass? Imaginary particles that travel faster than light would conceivably have such characteristics and are dubbed tachyons. These faster than light particles would have the extraordinary attribute that when they lose energy they go *faster*. They would be virtually undetectable because their velocity is faster than the photons that we would employ to perceive them. Physicists Herb Fried from Brown University and Yves Gabellini from the Université de Nice have proposed that dark energy is actually composed of tachyons. If this were indeed true, more than 68% of the energy in the universe is currently travelling faster than light. These tachyons would inhabit an upside-down universe were the speed of light is the lower boundary of velocity and it would require near infinite amounts of energy to slow an object down to that light speed barrier. The more energy the tachyon loses, the faster it would travel. Zero energy levels would cause near infinite speed. Is this tachyonic world the hyperspace that comic book movies refer to? Not exactly. When objects that move faster than the speed of light are plotted on a

Minkowsky diagram (a mathematical device used to relate two inertial reference frames at different velocities) they would appear to travel backwards in time to someone in an inertial reference frame that does not exceed the speed of light. This of course would violate causality.

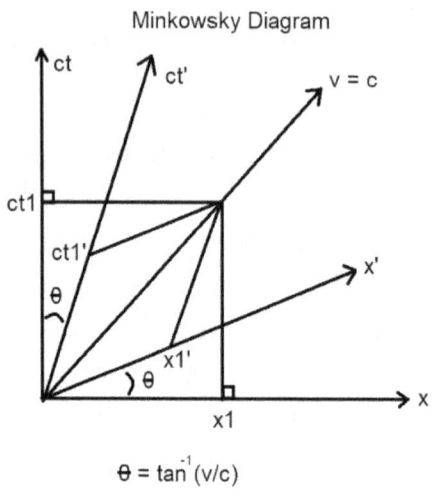

Minkowsky Diagram

$\theta = \tan^{-1}(v/c)$

(Primed inertial reference frame is traveling at velocity v)

One possible way to avoid violating causality is not to travel faster than light at all, but rather take a short cut through spacetime. This would use what is commonly known as a wormhole or an Einstein-Rosen Bridge. This idea was first proposed by Albert Einstein and Nathan Rosen in 1935.

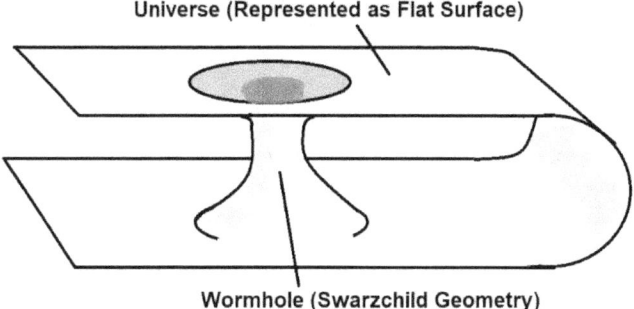

Universe (Represented as Flat Surface)

Wormhole (Swarzchild Geometry)

The Einstein-Rosen Bridge represents a shortcut (or tunnel) through space time caused by two locations of highly curved space somehow becoming connected. Unfortunately finding a naturally occurring wormhole would be problematic and finding one that goes somewhere useful even more problematic. Another difficulty is the wormhole would be inherently unstable. According to Professor Matt Visser at Victoria University of Wellington, quantities of something called exotic matter would be required to keep a wormhole from closing. Precisely what this exotic matter actually consists of is a matter for debate. However, it must possess something called negative energy to counteract the wormholes natural tendency to constrict (which would destroy anything trying to transverse it).

Proposed Warp Drive Ship

If wormholes are impractical, is it possible to create your own? Perhaps it is possible to bend space or even create a bubble in it that allows a traveler to circumvent the laws of physics. This of course is the famous warp drive from the Star Trek television series, or perhaps the hyperdrive of the Star Wars movies. Surprisingly enough, that is what was proposed by Mexican theoretical physicist Miguel Alcubierre in 1994. The proposed propulsion system would generate a warp bubble (actually a Lorentzian manifold) around the spacecraft. This bubble (more accurately a warp ring) changes the geometry of the space immediately surrounding the ship and creates a wave that contracts the space in front of the spacecraft and expands the space behind it. The ship does not move within this distortion, but rather is carried along as the region itself moves due to the operation of the Alcubierre drive.

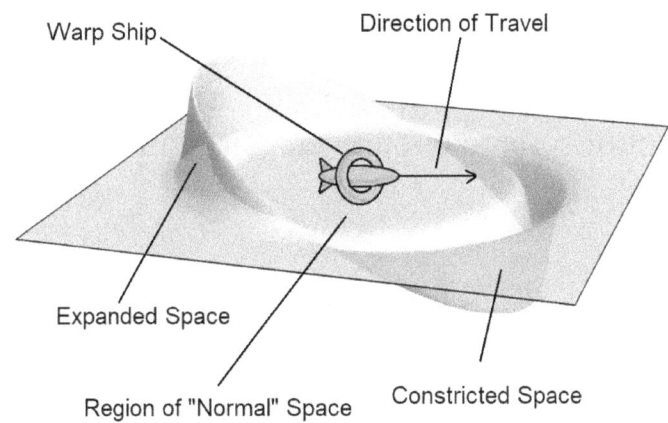

Alcubierre Warp Ring Spacetime Geometry

Unfortunately, the Alcubierre warp ship also requires exotic matter to function and would need a tremendous amount of energy. The original estimate to the energy was the mass of the planet Jupiter converted into pure energy (via the relativity equation). Later estimates after tweaking the drive functionality (such as changing the warp shape and oscillating it) have greatly reduced that original estimate. Another problem with the warp drive is as the ship travels through space, high energy particles would tend to get trapped in the warp bubble. These particles would gradually build up. When the warp field is disengaged, these particles would get released in the form of an ultra-high energy blast in front of the ship. On a long flight the blast could be powerful enough to vaporize entire planets (important safety note).

Another consideration for contemplating means of violating relativity is the idea of higher special dimensions. Currently M-Theory postulates 11 dimensions of space. Scientist speculate that the extra dimensions can't be perceived normally because they are "curled up" around the perceived three dimensions. Perhaps one of these higher dimensions could allow a short cut of sorts to circumvent Einstein's speed limits.

Higher Dimensional Space

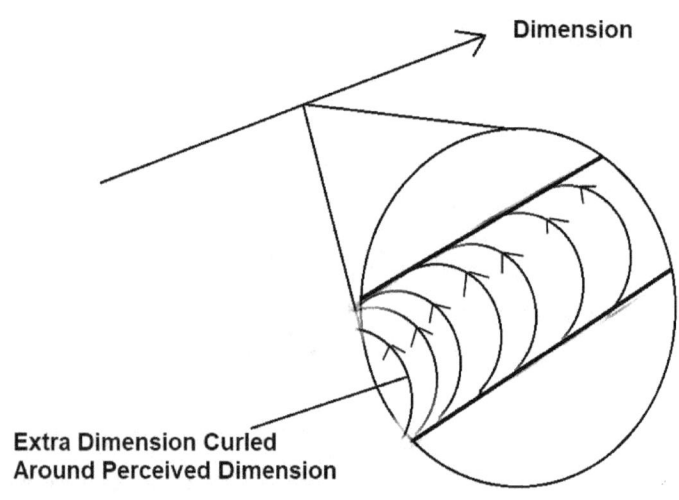

Another thought is: *what if the three dimensions we perceive are actually the curled-up ones?* What we experience as traveling in a straight line may actually be the equivalent of traveling around a tightly wound spring. If we were able to jump to a "non-curled" dimension, traveling a short distance along it would be equivalent to many times that distance in the curled-up dimension. We would have the effect of exceeding the light speed limit in one dimension while staying below it in another.

The strangest idea of all for an FTL drive involves something known as simulation theory (see Hacking the Universe). Simulation theory suggests that we are actually living in a simulation in a giant cosmic computer. If this actually is the case, it might be possible to get around light speed by somehow hacking the program. A rather unnerving thought.

Or perhaps faster than light travel will make use of some physical property or properties that we are as of yet unaware of. The number of things we don't know about the universe is likely far greater than what we currently know. Humanity has a history of overcoming obstacles. The greatest obstacle that we must overcome is not the lightspeed barrier, it is the barriers we create for ourselves.

The Distributed Mind

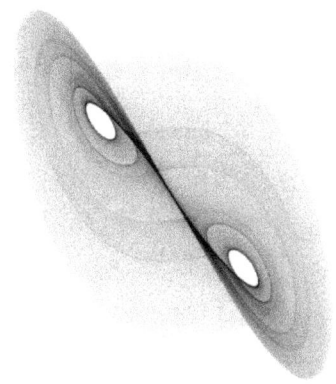

In the far future, a time where longevity and interstellar space travel (maybe even intergalactic travel) has become common place, how might these changes impact human life?

Well, if we have material needs taken care of, and lifespans are measured in the hundreds, or thousands, of years, and transportation is either instantaneous (teleportation) or so quick (a universal jump drive) it's effectively the same, then how might humans be affected by these changes? What will we look like?

My answer: the same as we do now – on the outside, that is. Oh, there'll be a fad of hotgevity, as soon as we figure out how to look youthful, after having spent decades watching our bodies decay. But eventually that will disappear. The concerns among the long-lived will turn to safety. Some have said that even with longevity the chances of a fatal accident will limit the average lifespan to sixteen hundred years give or take five hundred years or so.

So, what to do? To start some will research ways to minimize dangers. And in that process, some may turn to the past, to mythology for a clue.

There is one creature that exists in every cultural mythology. Angels, or their equivalent. And one point about angels: they don't die. At least I've never read an account in mythology, or theology, where one has.

Why? A religious person might argue because God created them that way. Another explanation might be that they are not a single contiguous physical entity.

In other words, if you kill such a creature you've only in effect cut off their fingernail. Most of their bodies are secluded elsewhere in places of power or secrecy. Hideaways where they place the controlling part of their essence, their central processing unit, as it were. Then they create bodies, which serve as automated arms and legs.

DISTRIBUTED CONSCIOUSNESS

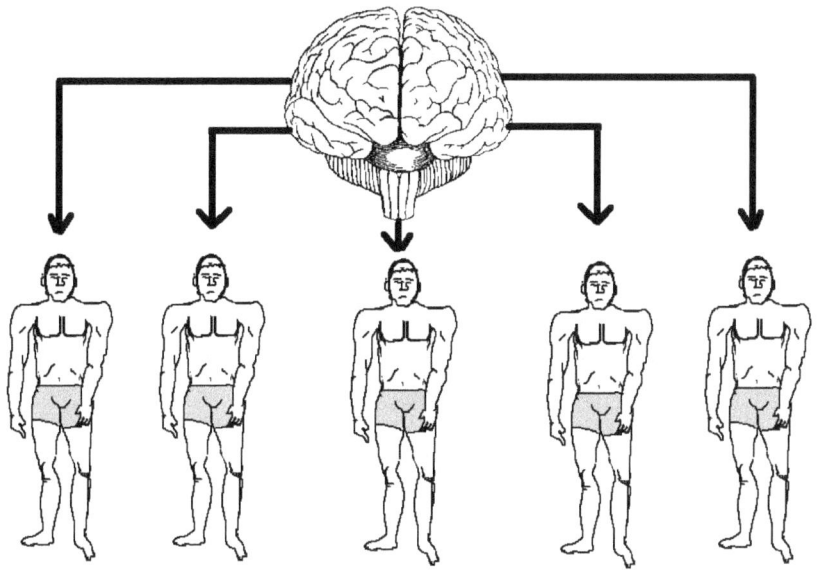

But how could such a distributed creature exist in reality?

Well quantum entanglement may allow for such a possibility. Action-at-a-distance's interference effect is superluminal. But we do not know how far the effect functions. But if it is instantaneous then maybe a distributed human mind and/or body is possible.

First, let us look at Feynman's Diagrams. They may intuit a way to investigate our mind's capacity to be spread sub-atomically across both space and time (see the Nature of Nature chapter). Then the Transactional Interpretation provides both a mathematical and a physical structure to explain the entangled nature of the non-locality of the action-at-a-distance effect.

It is these insights that may lead to an understanding of how the mind could be structured in such a way as to provide a non-localized conformation, or a distributed structure. Then with that structure such a mind could operate separate bodies in much the same way that we operate different arms, legs, in coordination with our five basic senses. Thus, if one body is destroyed, the rest of the organism survives, even the experiences, and the consciousness of that body could rapidly move to the greater whole and no loss in consciousness occurs.

This is not just longevity; it is almost virtual immortality.

How would such a human be different from us? Well, while we may be thinking of dozens of items at once, it is difficult for even the best multi-tasker to do more than several things simultaneously with any degree of competence.

Such a distributed, or multiplex person could do many things in many different places. Some tasks that may require intense concentration over not just days, or months, or years, or decades, but maybe centuries or millennia could be done. Then other activities which require cooperation of multiple bodies from the same distributed mind could be attempted. Imagine how such an existence would in some ways radically, fundamentally alter the nature of what it means to be human?

Such an entity would prove a formidable challenge to today's concepts of humanity. Everything from theology to psychology to sociology to biology would be up for grabs.

Some observers contemplating the far future and seeking to immanentize the eschaton have looked at ways of how theology transforms or transcends human nature. Indeed, it is the transformative nature of many major religions that are their selling point: the change in the twinkling of an eye, as it were.

Well, what would such a transcendentally, transformed human look like? How would they be different than a distributed human who had achieved such capabilities through a secular process?

I can think of only one difference between a secular, distributed far future human, and a theologically altered human being (who could be given all of the above distributed abilities): the propensity to sin; to do evil. And make no mistake such a distributed human could do evil on an unimaginably vast scale.

The promise of divine redemption on one side, and ultimate individual power on the other. Both require awesome responsibility to handle such power in a decent fashion. For the secular creature (to call such a distributed entity a human almost seems a misnomer) the responsibility lies within. For the creature transformed by God the responsibility is jointly shared, although the power, if not the responsibility, ultimately comes from without.

So, how do you view such possibilities? These are concerns that once were the sole province of pastors and prophets, of priestly potentates and monkish philosophers. Well, now they are the concern of everyone, as they should be.

The authors of this book believe that such considerations are best considered if we have literary scouts looking ahead, boldly going to Canaan and returning to report to us of the giants living in that land. It is up to us if we want to be bold and voyage into that new land, or turn our backs like many of the ancient Hebrews, or like many of the Science Fiction writers at the 1994 Nebula Awards, and slink from the challenge of experiencing through literature all of the possibilities of an Elegant Future.

You see an Elegant Future does not mean the end of high drama and adventure, but merely the beginning.

Hacking the Universe

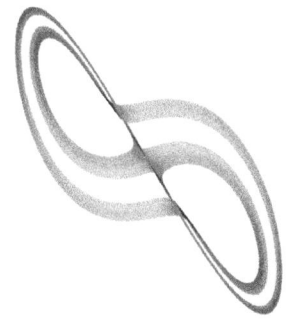

Did you ever get the feeling that there was something wrong with the universe? Feelings of Déjà vu or impossible coincidences? Do the events of the day seem to happen in familiar patterns? Do you wonder if you're still asleep when you're actually awake? You're not the only one with concerns about the nature of the world. People have had misgivings about the universe for centuries. As far back as 1641 AD French Mathematician and Philosopher Rene Descartes wondered about reality. He is quoted as saying: *"I suppose therefore that all things I see are illusions; I believe that nothing has ever existed of everything my lying memory tells me. I think I have no senses. I believe that body, shape, extension, motion, and location are functions. What is there then that can be taken as true? Perhaps only this one thing, that nothing at all is certain."*

The detailed study of the universe has served to reinforce rather than eliminate our doubts about its nature. The study of light, the phenomenon that is vital to our perception of the world we exist in, led to the disturbing wave particle duality paradox. Further study of this peculiarity yielded the Bell inequality and the idea of entanglement. More experiments in this bizarre quantum realm lead to the mind-bending results that indicate observations can actually affect reality and change experimental outcomes. These experimental results altered the tangible, straightforward, and mechanistic view of the universe into a more ghostly and surrealistic assessment. The feeling that the universe is not exactly what it seems to be grows stronger the more we study it. Even the study of empty space has led to the conclusion the emptiness (or hard vacuum) is not really empty at all but filled with a foam of probabilities as proposed by John Archibald Wheeler in 1955.

In 1993 Gerard 't Hooft proposed a theory known as the Holographic Principle. Later in 1995 Leonard Susskind crafted a precise string theory interpretation of the holographic principle. This principal suggests that the universe only appears to be three dimensional, but rather is actually a hologram (three-dimensional information encoded into a two-dimensional surface). The three dimensions we perceive are actually an illusion. A more accurate stating of the holographic principal is that *all of the information enclosed within a region of space can be reconstructed from the information on the containing surface.* The idea for this weird concept was obtained by the study of black holes and the idea the information about objects that fell into the singularities could not be destroyed, so it must be stored on the event horizon instead.

As strange as the holographic principal seems, things get even stranger. In 2003, Nick Bostrom of Oxford University proposed "the simulation argument." In the simulation argument, Bostrom theorized that humans are actually living in a *post-human* computer simulation of the universe. This idea became reinforced when Dr. James Gates, a theoretical physicist at the University of Maryland, discovered a set of mathematical equations in his study of string theory that appear to be indistinguishable from a form of computer code. He also identified what appears to be error-correcting code in the recursive mathematical expressions in the supersymmetry equations used in string theory. This forced Gates to wonder if we were all actually living in the Matrix.

Many scientists are becoming more and more convinced that the universe acts in many ways like a computer program. Physicist Stephan Wolfram's studies of cellular automata discovered that very simple programs can produce extremely complicated results. Wolfram calls the idea that complex structures can be generated by simple computational processes and that different complex forms can be compared by comparing the underlying computations required to generate them the *Principle of Computational Equivalence.* Many argue that Wolfram's notions are too vague and abstract, but nobody can refute the fact that simple expressions can spawn very complex outcomes.

The generation of complexity from the simple is also seen in the study of fractals, where very simple mathematical constructs generate unbelievably intricate and beautiful patterns. It seems the simplest rules can produce irreducible complexity. Perhaps the vast complexity we see in reality is the result of very simple computations occurring in the cosmic computer. What would this cosmic computer be? Whose desk is it sitting on?

A Fractal – Complexity from the Simple

Seth Lloyd, professor of mechanical engineering and physics at the Massachusetts Institute of Technology, thinks the Universe is a gigantic quantum computer. According to Professor Lloyd, all particle interactions in the universe exchange not just energy, but also information. Elementary particles not only collide, they also compute. Lloyd gets his ideas from his study of quantum computers which use quantum interactions to process information. The basic unit of a quantum computer is not a bit, but rather a qubit. The qubit is a two-state (or two-level) quantum-mechanical unit, which behaves with the peculiarity of quantum mechanics. The measured result of a qubit is still either a one or a zero, but quantum mechanics allows the qubit to be in a superposition of both states/levels simultaneously until it is measured.

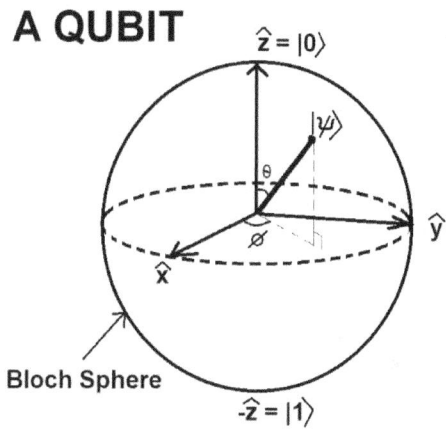

A QUBIT

(The Qubit collapses to zero or one when measured.)

If the universe is a program executing in a giant computer (quantum or otherwise) what are the ramifications of this? Is this computer conscious? Who or what created it (or did it create itself)? Most importantly, can we use this concept to our benefit? Can this computational reality program be hacked? If we even attempt an act of corporeal program meddling, will the cosmic cyber police show up at our doorstep with a warrant for such a crime?

COMPUTER ARCHITECTURE

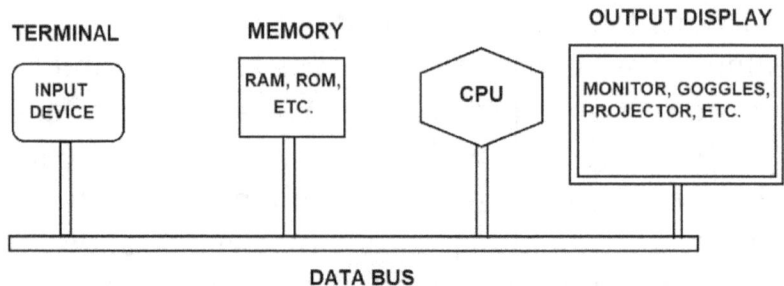

How could we even begin to hack the program that generates the world? If the universe is some sort of a computer, then we must examine a computer and compare it with reality. The computer has several basic components, one or more central processing units, memory to store data, an input terminal, and an output device (either a monitor, or projector, etc.). What would be the analogs for the universe itself? One would need to find the input device for the universe, or it may have to be constructed. To build a cosmic terminal, we would need to understand how to connect to the data bus for the cosmos. This may be difficult to do if we are inside the program itself.

Where is the data bus of the universe? Where does the low-level code operate? Perhaps this is an explanation for entanglement and the Bell Inequality. Common belief holds you cannot send information using entanglement, because as soon as the quantum state of the entangled pair change occurs, the entanglement ceases. Perhaps entanglement is a temporary connection using the data bus of the universe. Or maybe entanglement occurs when two particles temporarily share the same memory location. In programming terms, they would be pointing to the same address. The pointing would be via this cosmic computer bus.

One can draw this analogy even further by examining object-oriented programming (OOP). An object in the world of OOP has two types of components, attributes and methods. Attributes describe the object. Methods cause the object to do things. There can be multiple instantiations of an object if the object is dynamic. There is only one instantiation of a static object and the static object's methods can be called without creating an instance of it. Particles would be dynamic objects. Fundamental forces would be static objects. Both types of objects work because of the universal computer bus, and the entanglement phenomena could be the direct effect of the actions of that bus in the universal program. Weird stuff if true.

One example of this is position. If position is simply an attribute of an object, then perhaps it could be instantly changed to a new value. The object would vanish and instantly appear in another position. This would violate relativity, but perhaps the speed of light limit is a result of the computational speed of the universal CPU when you change position by incrementing its coordinate values. Incrementing a memory location requires running the program in a loop and adding or subtracting an incremental amount to the location until it reaches the required value. This would take a long time for very large values. The faster you try to run the increment; the more energy is required. Changing a value to a new value directly would take only one cycle and virtually no time at all.

Is the universe a computer, or is it a mind? At a certain point is there any difference? Perhaps the difference between a mind and a computer is simply the existence of consciousness. If the universe is a mind, then hacking it has moral implications. We can, for now, take some solace in the fact that we currently have no clue if this is even possible, or how we could accomplish this. However, mankind is a curious creature. More experiments lie ahead, and more secrets will become revealed. Hopefully this knowledge can be used elegantly for the good of all, rather than maliciously for destruction.

PART 5:

DETERING MAN'S WORST ENEMY: HIMSELF

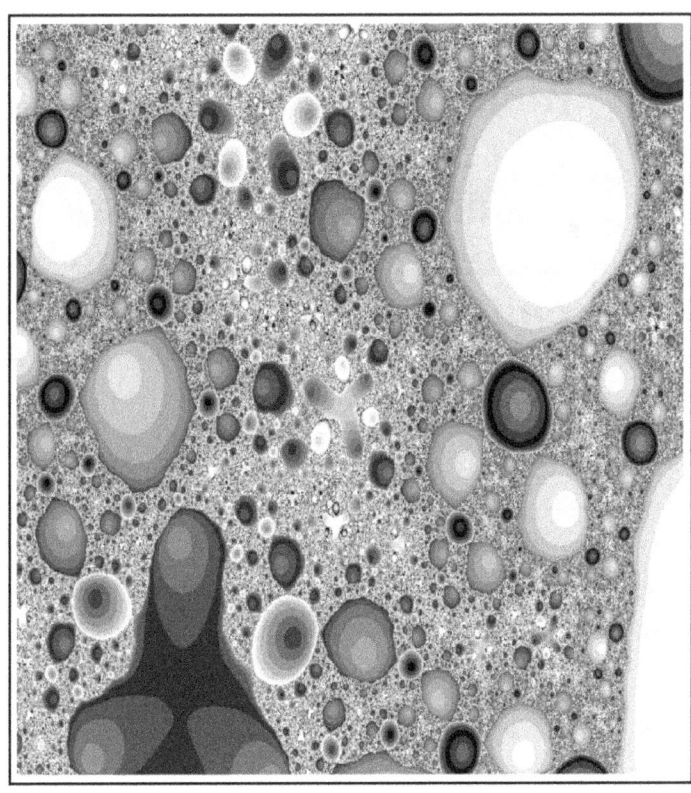

Every great dream begins with a dreamer, Always remember, you have within you the strength, the patience, and the passion to reach for the stars to change the world.

-- Harreit Tubman

The Ownership Society

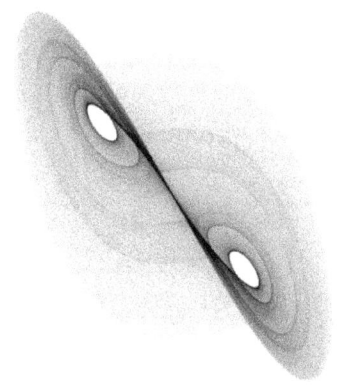

"Those who do not understand history are doomed to repeat it." – George Santayana, 1948.

Beginning centuries before those words were spoken humanity witnessed the resource rape of two continents. Galleons full of gold and silver bullion sailed every day from the so-called New World of the Americas South and North back to Europe – Spain specifically – to fill the coffers of the ruling dynasty of that country: the Hapsburgs (rulers at that time of not only Spain, but at times of her Iberian sister country Portugal, the Low Countries of Holland and Flanders, the Kingdom of Naples – most of the Italian peninsula south of Rome, and Austria). Imagine all of the easily available wealth of nearly a quarter of the Earth's surface siphoned off by one in-bred European royal house.

The Elegant Future

This was theft on an unimaginable scale. Theft, impure, but simple.

In today's world of social justice warriors where is the outrage? But that was the mediaeval world of monarchies.

That was then.

Where the inhabitants of the Americas had no choice as they watched their heritage being stolen and taken from the ports of Cartagena and Veracruz to Seville and on to Madrid, there to be used by incest-challenged royals.

Then came a new nation whose history has never been stained by royalty. A new nation with a new creed.

Life, Liberty and the pursuit of Property.

At least that was the John Locke quote from Thomas Jefferson's first draft of the Declaration of Independence before someone (many accounts credit Benjamin Franklin) suggested the now familiar quote of life, liberty, and the pursuit of happiness. Imagine how differently we would view property -- or the ownership of property – if that had been in the founding document of our country?

And beyond social questions of equality there are economic ramifications. Would we have the great disparities of living conditions that we have today if we had viewed ownership differently? From great mansions in gated communities, or baronial country estates to homelessness, poverty, countless tens of thousands living in hovels in the woods, or under underpasses, the disparity of living conditions is an intolerable disgrace for the richest, most powerful country in the world.

One thing is certain in order for an Elegant Future to happen we cannot have great disparities of living situations, or opportunities. Automation will continue to hollow out employment, and thus, economic opportunities. Continuing income disparities will then inevitably rise to literally astronomical levels in the future.

So, what can be done? How can we prevent a future where America and the entire first world looks like a third world shamble such as Rio de Janeiro, with phenomenal wealth cheek-to-jowl with sprawling shantytowns of corrugated roofs and cardboard walls with no insulation, unsanitary conditions, certainly no privacy, and ultimately no dignity?

The answer is right over our heads. The sky. The Infinite Frontier as Robert Zubrin has called it. It opens up at night and shows us the blazing stars of the Milky Way Galaxy, a few hazy patches of light, other galaxies, the Mighty Andromeda, and a few moving objects, called by early astronomers Wanderers, which we today call planets, and a few even smaller objects – objects of great interest we call the asteroids.

The crucial concept is this. Asteroids will usher in an era of extreme human wealth. Yes, there will be unforeseen consequences. During the mineral wealth transfer to Spain from the Americas a side effect was the collapse of the European silver market, which dealt a financial blow to the Ottoman Empire from which they never recovered.

But incredible fortunes will be created. Wealth on a heretofore unfathomable scale will be generated and shifted around. What we need to make sure of is that we do not end up with a few trillionaires, or quadrillionaires, with a vast, ever increasing population of huddled, hungry, shivering homeless persons.

How to accomplish this? Well the key to unlocking a wealthy future for nearly all humans rests on considering the potential of a document, The Moon Agreement (1979), the culmination of five treaties dating from 1960s and 1970s, done under the auspices of the ad hoc Committee on the Peaceful Uses of Outer Space (COPUOUS) created in 1958 by the United Nations General Assembly.

The Moon Agreement refers to all "celestial bodies in the solar system, other than the Earth" and objects that fall into the Earth's atmosphere (meteorites). It exempts scientific and exploratory activities (getting soil samples, etc.), but on economic activities, other than the vague statement that the Moon and other celestial bodies are "the common heritage of mankind," the document is silent. Neither does it mention them (mining the asteroids, for instance), nor prohibit them in specific language. It also does not mention, and I doubt its designers even considered, radiation from any such object, the Sun in particular. So, the collection of solar power in space and its commercial exploitation is not regulated in the Moon Agreement.

Now, the practicality of this Agreement is moot, one might even say nugatory and moot. Most of the nations involved in early space exploration are not signatories. And the language is too vague and the definitions too imprecise to carry any legal weight beyond a debating exercise.

So how to use these concepts to alleviate human suffering while preventing an oligarchical economic takeover of the solar system's wealth?

Well, one way is for those who intend to mine the asteroids to agree to the concept of shared ownership of the solar system. Ownership in which those using "celestial bodies" (the asteroids and moons, etc.) agree to pay a royalty to humanity for their use. Say ten percent.

So, a valuable asteroid (some estimates place the wealth of the mineral contents of certain asteroids in the quintillions of US Dollars) could – even with the devaluation caused by supply and demand – return millions of dollars to every human. Now considering the numbers of asteroids that could be mined, the amount of royalty due to each human is spectacular.

Before we begin with the objections let us note that there is a precedent for this concept in action covered by American law. The State of Alaska offers a variant of this concept to their citizens in the form of compensation for oil extraction.

There are many obstacles to such a plan. Most entrepreneurs would likely object to any intrusion on their profits. Most major space-faring nations would object. They have neither signed the 1979 Agreement nor are they likely to sign any future agreement.

Nations are problematic, but as for the entrepreneurs well Warren Buffett has convinced many billionaires to put half of their wealth into philanthropic activities, charitable foundations, et al. Here, we are talking ten percent. Sure, there will be costs, and overhead, and taxes, but ten percent? How much of a trillion-dollar haul will make them happy?

Do they want to live in a society of a super-wealthy few, living in gated communities surrounded by vast hordes of the homeless and other folk barely existing on meager handouts? Or would they rather live in a society where everyone is wealthy by early 21st Century standards? Where they are super-wealthy, but do not stand out to the point where they need bodyguards and walls to protect them from the starving masses?

Given that the above arguments somehow hold, then the structure of the disbursement of such royalties will have to be carefully structured. Fees should be minimal. The organization overseeing this, COPUOUS, or some variant thereof, should be subjected to strict oversight. Accounts should be set up with banks who agree to disburse the funds for a pre-agreed minimal fee. With the internet we have the structure to download the money into every person's private account.

Santayana can be disproved, at least for this once.

Imagine a child of the 22nd Century, who lives on the astrolith of a repurposed asteroid where after mining has been turned over to sustainable agriculture? Each day she tends to the produce grown and then sent via teleport to other humans throughout the solar system. Fresh food for everybody. A future where everybody has enough funds for their needs.

And for her the future is bright. The first interstellar flights are underway. She can be whatever, or whomsoever she wants to be.

And the injustices of a half a millennium ago are forgotten and the Empire of the Hapsburgs a mere footnote in history.

Now, that's an Elegant Future!

END NOTE:

ETE: Astrolith is a word I created by combining a variant of the first prefix of asteroid with the suffix of the word regolith (the term for lunar soil, as in lunar regolith). So, astrolith is the term for soil on an asteroid.

The Elegant Future

Extreme Wealth: What Happens When Everyone is Rich?

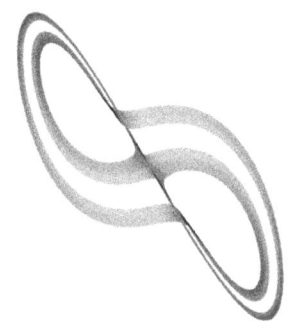

There are currently more than 7,000,000,000 human souls scuttling about the surface of this pale blue speck we call the Earth. They are, for the most part, trying to make the best of their lives. They are working at jobs, trying to raise families, following their daily patterns, and looking for ways to make their tedious lives just a little bit easier. A large percentage of them classify among the poor. A smaller percentage of them are homeless. These unfortunate ones are living on the streets. They are not certain where their next meal is coming from. Some suffer terribly, looking through garbage cans for their next meal. Some reside in tents along the freeways. Others live in cardboard boxes in back alleys or under bridges. Some of the ones who are the most destitute suffer from mental illness or drug addictions that make them nearly impossible to employ due to their unpredictably chaotic personalities. Some are there by choice, unable to adapt to the daily schedule that a job demands. But does it really have to be this way? Those who are mentally ill may need dramatic intervention or institutionalization, mostly to protect them from themselves. For the vast majority, all they really need is a permanent residence. It's hard to apply for a job if you don't have a mailing address. Will it ever be possible to help everyone who needs and wants help?

To accomplish this comes down to sufficient resources. One might consider the idea that raising taxes on the fortunate is the only way to aid the unfortunate. Punishing people who are financially successful by imposing large taxes on them is not a good way to gather resources. Gaining the needed funds this way destroys incentive and will eventually depress the economy and make matters worse. If you punish success you will end up with nothing but failure. Help should always be considered a gift, not an obligation. Giving out assistance without the unspoken agreement that people must at least try to learn to create resources for themselves makes the assistance the end and not the means. This becomes a trap for those of limited willpower. More resources are required to be able to pay.

Currently there are simply not enough resources available to solve the financial dilemmas that currently face the world. But there is a way to change that. If we look a little beyond the boundaries of the Earth, there are vast resources that could change everything. The Moon, the asteroids and other celestial bodies are abundant with raw material that could generate almost limitless wealth.

The Earth's resources are finite. One example of this is the precious metal gold. The U.S. Geological Survey estimates that there are only about 57,000 tons of gold left on the Earth to be mined. The resources in space are almost limitless. Just one asteroid like Psych 16 is estimated to contain 400,000,000 tons of gold, and that's just the gold. That's 7,000 times as much gold as exists on the Earth from just one asteroid. And there are millions of asteroids larger than 1 kilometer in diameter. The resources in space are so vast the calculations become ridiculous.

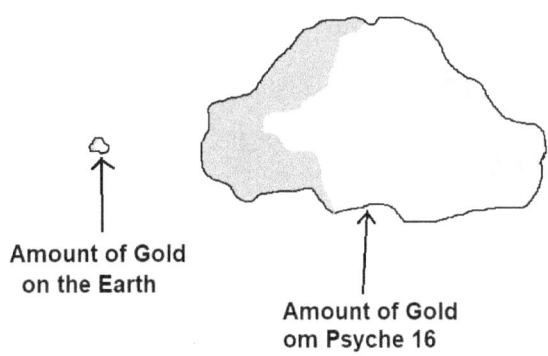

Amount of Gold
on the Earth

Amount of Gold
om Psyche 16

Space resources, if properly utilized, have the potential to make everyone on Earth rich, but is it actually possible for everyone to be rich? Or will, as past history has shown, all the wealth go to a few? Despite the existence of vast wealth disparities in today's world the standard of living has increased remarkably in the last two hundred years. The resources from space could accelerate this process and assure that everyone on this planet is pulled from hopelessness and misery. They could provide everyone with at least an equal opportunity to succeed. This can be done without having to punish success. As the asteroids jointly belong to everyone, even the tiniest royalty for the right to access the vast resources orbiting the Sun would yield an immense amount available for philanthropic endeavors. With most of the labor being carried out by automation, humans would be free to pursuit more enlightened professions such as becoming craftsman, artisans, designers, and researchers. Automation would perform the more mundane and repetitive tasks, allowing humans to fill jobs requiring decision making, or tasks requiring more skill, contemplation, insight, or just plain imagination.

What would such a society look like? It's hard to view past the singularity of unlimited resources and wealth. We see some hints occurring today. People may choose to take voluntary risks to seek sensory thrills, and possible spiritual enlightenment, as we see today with hang gliding, rock climbing, or other risky endeavors that stretch the boundaries of human experience.

Others may decide to research new unknown areas, as those currently resourceful souls that are building fusion reactors in their basements. The larger our knowledge of the world grows, the greater the circle of the unknown becomes. There will always be new things to discover.

One can also hope that many will find meaning in helping their fellow man. Searching the world for others that have fallen through the cracks of life, and rescuing people from their circumstances or themselves. There is no greater reward than to give those who are in need a helping hand, or at least a few words of encouragement when they have lost their way.

If space travel becomes easily available to the masses, many will set off into the great unknown. The universe is, for all extents and purposes, endless, and thus is a source of unlimited intrigue and adventure. Just pick a direction and off you go. Unlimited lifespans will greatly help such explorers as they venture out into the infinite frontier.

Others may resolve to explore the spiritual side of reality, exploring the boundaries of consciousness itself. The human mind is a vast wilderness and has yet to be fully explored. The choices are as endless and as elegant as the universe itself. The future holds far too many possibilities to be anything other than optimistic about what lies ahead.

The Elegant Future

Afterword

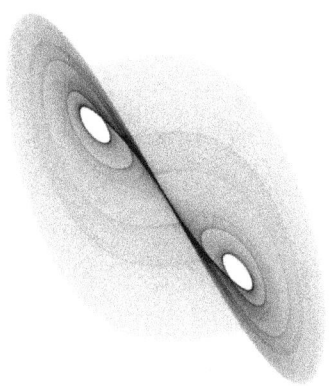

It has been my fortune to know many smart and decent men. You see, smartness alone is never enough. What we need for the future to become elegant is a combination of both of these traits. Which is why this book discusses income inequality and bullying, or as I term it: the culture of cruelty. We must become BOTH smarter and better.

And part of that is doing what Gene Roddenberry did so beautifully in the concept of Star Trek. Showing a future where humanity not only survives, but becomes better by overcoming social problems, such as poverty and hunger and racism. It gave us not only hope, but guideposts on how to be better as a society, as well as be better individually.

The Elegant Future

When I saw the field retreating into dystopic despair, I felt that some statements needed to be made. We all need warnings from time to time, but all dystopia all the time (the Mad Maxing of the future, as it were) was too much.

I mentioned this concern to my longtime friend and colleague, Craig Philip Peterson, and he agreed and suggested the book that you are now reading, and hopefully, enjoying. We decided to put together a book that could show how fantastic the future could be if we, as a society, took on social concerns and technical concerns together.

I was delighted at the depth and brilliance of Craig's writing and thinking. Not surprised though, because I have been fortunate enough to know him from when we were both students (we met at a science fiction club when I was matriculating at Willamette University and he was completing his secondary education at South Salem High School) some forty plus years ago. And from the beginning, I have been amazed at the brilliance and originality of his mind and the renaissance breadth of both his creativity and his interests.

Since then I have met many other brilliant men, but one person stands out for not only his towering accomplishments in our understanding of the basic structure of the universe, but his desire and his ability to convey that to the interested layman. His name is Dr. John G. Cramer. He is a Professor Emeritus of Physics at the University of Washington, located in

Seattle. His Transactional Interpretation of Quantum Mechanics not only explains what happens at the collapse of the wave function, but also explains in brilliant mathematical and physical models the mechanisms behind non-locality, or what Albert Einstein termed "spooky action at a distance."

When I began attending the annual Seattle Science Fiction convention, Norwescon, I became aware of Dr. Cramer's lectures and panels on physics and other, mostly, science-related topics. Besides being a ground-breaking theorist and experimentalist, he is a great lecturer and a profoundly brilliant teacher. I took notes during his lectures and panels and although it took years for what he was presenting to sink in, there came a day when quantum mechanics and quantum entanglement not only made sense, but I could understand their underlying elegance and beauty.

For anyone willing to listen and learn, his willingness to share his knowledge was, and is, a tremendous gift to everyone fortunate enough to hear his lectures and read his books and his column in Analog Magazine. I have never thanked him properly in private for sharing his knowledge and wisdom, so this epilogue shall have to suffice: thank you, Professor Cramer for providing me and anyone who attended your panels and lectures for decades of enlightenment.

Needless to say, I am delighted to share this book with both of these talented and brilliant men. I am proud to call them friends, and even prouder that they are not only brilliant, but also decent in the most profound sense, in that both of them want only the best for people both individually and socially. And nowhere is this more obvious than the passion that both of them share for longevity. For the sad reality of current human existence is that just as we as individuals begin to develop wisdom and understanding we die. It is thus in everybody's interest, young and old, that longevity happens.

And that would be the beginning of creating an Elegant Future.

Elton Elliott

About the Authors

Craig Philip Peterson is currently working as a software engineering consultant in the greater Seattle area. He has held many diverse occupations including automobile mechanic, independent filmmaker, actor, chemist, electrical engineer, and software development engineer. He has always been interested in science, science fiction, and fantasy and has a large collection of books devoted to those subjects. He is the author of THE ADVENTURES OF JONATHON FARMER series, and co-author of THE INFINITY ANOMALY, and CATACOMBS OF DOOM. Craig loves to draw and is responsible for all of the interior illustrations and the cover art. He holds degrees in physics and electrical engineering and lives in Washington State with his wife and son.

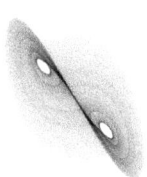

Elton Elliott is a Northwest author, editor, and publisher. His most recent projects include co-authoring the novels PRINCE OF EUROPE and BISHOP OF ROME, and the short fiction piece A Quantum Field of Ghosts and Shadows. His short story, Space Aliens Taught My Dog to Knit, (written with Jerry Oltion) appeared in Analog magazine. A former editor and publisher of Science Fiction Review, his mass market anthology NANODREAMS (Baen Books, 1995) was reviewed in Scientific American (April 1996).

www.ingramcontent.com/pod-product-compliance
Lightning Source LLC
Chambersburg PA
CBHW060825170526
45158CB00001B/81